Mohamed Mekewi
Haidy Kamal

Nano ZnO antimicrobial additives to epoxy and Acrylic paints

LAP LAMBERT Academic Publishing

Impressum / Imprint

Bibliografische Information der Deutschen Nationalbibliothek: Die Deutsche Nationalbibliothek verzeichnet diese Publikation in der Deutschen Nationalbibliografie; detaillierte bibliografische Daten sind im Internet über http://dnb.d-nb.de abrufbar.
Alle in diesem Buch genannten Marken und Produktnamen unterliegen warenzeichen-, marken- oder patentrechtlichem Schutz bzw. sind Warenzeichen oder eingetragene Warenzeichen der jeweiligen Inhaber. Die Wiedergabe von Marken, Produktnamen, Gebrauchsnamen, Handelsnamen, Warenbezeichnungen u.s.w. in diesem Werk berechtigt auch ohne besondere Kennzeichnung nicht zu der Annahme, dass solche Namen im Sinne der Warenzeichen- und Markenschutzgesetzgebung als frei zu betrachten wären und daher von jedermann benutzt werden dürften.

Bibliographic information published by the Deutsche Nationalbibliothek: The Deutsche Nationalbibliothek lists this publication in the Deutsche Nationalbibliografie; detailed bibliographic data are available in the Internet at http://dnb.d-nb.de.
Any brand names and product names mentioned in this book are subject to trademark, brand or patent protection and are trademarks or registered trademarks of their respective holders. The use of brand names, product names, common names, trade names, product descriptions etc. even without a particular marking in this work is in no way to be construed to mean that such names may be regarded as unrestricted in respect of trademark and brand protection legislation and could thus be used by anyone.

Coverbild / Cover image: www.ingimage.com

Verlag / Publisher:
LAP LAMBERT Academic Publishing
ist ein Imprint der / is a trademark of
OmniScriptum GmbH & Co. KG
Heinrich-Böcking-Str. 6-8, 66121 Saarbrücken, Deutschland / Germany
Email: info@lap-publishing.com

Herstellung: siehe letzte Seite /
Printed at: see last page
ISBN: 978-3-659-66168-6

Zugl. / Approved by: Ain Shams University, 2014. Egypt

Copyright © 2014 OmniScriptum GmbH & Co. KG
Alle Rechte vorbehalten. / All rights reserved. Saarbrücken 2014

Mohamed Mekewi
Haidy Kamal

Nano ZnO antimicrobial additives to epoxy and Acrylic paints

Enhancement Activity of Antimicrobial Additives to Epoxy and Acrylic Emulsion Paints

By

Dr. Mohamed Ahmed Mekewi
Professor of Polymer and Material science
Faculty of Science, Ain Shams University, Cairo. Egypt

Dr. Maged Shafik Antonious
Professor of Physical Chemistry
Faculty of Science, Ain Shams University, Cairo. Egypt

Dr. Abd El-Fatah Mohsen Badawi
Professor of Petrochemicals
Egyptian Petroleum Research Institute, Nasr Cit, Cairo. Egypt

Dr. Adel Mohy El-Din Gabr
Head of R&D, Pachin for Paints, Obour city, Cairo

Dr. Khaled El-Baghdady
Assistant Professor of Microbiology
Faculty of Science, Ain Shams University, Cairo. Egypt

Haidy Bahig Kamal
Assistant Lecture of Physical Chemistry
Faculty of Science, Ain Shams University, Cairo. Egypt

2014

Enhancement Activity of Antimicrobial Additives to Epoxy And Acrylic Emulsion Paints

By

KAMAL, H. B[1], MEKEWI, M[1], ANTONIOUS, M S[1], BADAWI, A M[2], El BAGHDADY, K[3], GABR, A. M[4]

[1]Department of chemistry, Faculty of science, Ain-Shams University. Cairo, [2]Egyptian Petroleum Research Institute (EPRI), Nasr City, Cairo, [3]Microbiology department, Faculty of science, Ain-Shams University. Cairo, [4]Pachin for Paints, Obour city, Cairo.

Correspondence should be addressed to:
Dr. Mekewi, MA, e-mail address: mahikewi@yahoo.co.uk

Contents   Page

Part I
Imparting incremental properties to epoxy paints using decylamine silicate:$CoCl_2$ and mercury complex compound additive.

Abstract	3
Introduction	3
Materials and Methodology	4
1. Chemicals	4
2. Paint coat preparation specimens	4

3. Synthesis of paint additives
    - n-Decylamine silicate (DAS)
    - Bis-(diphenylthiocarbazono) mercury(II) complex (Hg $(HDz)_2$)
	4

4. Preparation of paint samples
    - Native epoxy paint
    - Epoxy paint added DAS:$CoCl_2$
    - Epoxy paint added Hg $(HDz)_2$
	6

5. Structural conformation	6
6. Physico-mechanical properties	6
7. Thermogravimetric analysis (TGA)	6
8. Chemical resistance	6
9. Photo Stability	7

10. Antimicrobial activity
    i. Bacterial clinical isolates
    ii. Growth and maintenance of bacterial clinical isolates
    iii. Preparation of bacterial inoculum
    iv. Preparation of paint discs
    v. Bacterial sensitivity test
    vi- Statistical analysis
	7

Results and discussion	9
1. FT-IR structural studies of epoxy paint and additives	9

- Epoxy paint and DAS:CoCl$_2$ additive
- Epoxy paint in presence of Hg (HDz)$_2$

2. Physico-mechanical properties	13
3. Thermogravimetric analysis (TGA)	14
4. Chemical resistance	17
5. Resistance to alkali and acid attack	18
6. Resistance to artificial sea water	20
7. Photo Stability	22
8. Anti-microbial activity examination	24

Part II
Nano ZnO/amine composites antimicrobial additives to Acrylic emulsion paints. — 28

Introduction	28
Materials and Methodology	29
1. Chemicals	29
2. Preparation and Synthesis	
a- Preparation of ZnO nanoparticles	30
b- Synthesis of nano ZnO composite	
3. Paint and additives: structural, thermal and antimicrobial features and characterization	31

    a- Chemical structure elucidation using FT-IR analysis
    b- Thermo-gravimetric Analysis (TGA)
    c- Transmission Electron Microscopy
    d- Antibacterial Activity
      i. Preparation of bacterial inoculums
      ii. Preparation of paint discs
      iii. Antibacterial activity of paint discs
        - Agar diffusion method
        - Turbidity method
      iv. Antimicrobial Data Statistical Analysis

Results and discussion	33
1. Chemical structural analysis (FT-IR)	33
2. Thermal stability (TGA)	35
3. Nano ZnO and ZnO/amine composites structural features (TEM)	36
4. Microbiological activity of native and antimicrobial added acrylic paints.	38

Part III
Characterization of the three member's additives as in-can preservatives. — 44

Conclusion	50
List of Tables	51
List of Figures	52
References	54

Part I
Imparting incremental properties to epoxy paints using decylamine silicate:CoCl$_2$ and mercury complex compound additives

Abstract

The aim of this work was to improve the characters of epoxy-based coatings to increase their stabilities, temperature tolerance and antimicrobial activities. This was carried out through the use of effective additives such as n-decylamine silicate:cobalt chloride (DAS:CoCl$_2$) and bis-(diphenylthiocarbazono)mercury(II) complex (Hg (HDz)$_2$). FT-IR structural studies, thermal gravimetric analysis, mechanical properties and resistance to acids, alkalis and artificial sea water were studied. Perpetuation of the epoxy paint major properties was optimized towards thermal properties improvements. The modified epoxy paints were able to inhibit some important pathogenic bacteria including MRSA, VRSA and *Pseudomonas aeruginosa* clinical isolates. As a result the paint could be applied in hospitals and surgical rooms to decrease contamination and spreading bacterial infections. Curing reaction mechanisms were offered in presence of either additives leading to understand the physico-mechanical and chemical properties upgrading and marked results.

Key words: Epoxy paint, n-Decylamine silicate:cobalt chloride , Bis-(diphenylthiocarbazono)mercury(II) complex and Curing reaction mechanism.

Introduction

The modern coating technology requires those of superior mechanical, thermal and anticorrosive characteristics to overcome the adverse environmental conditions. Currently new field of organic-inorganic hybrid polymers are now being tackled [1], including epoxy paint [2] due to its superior strength, low shrinkage, better bonding with different substrates, good dimensional stability and long term corrosion and chemical resistance [3]. Use of metal-containing epoxy resins allows the possibility of producing epoxy polymers with good mechanical properties and high thermal stability, as well as achieving low processing temperature including mercury based additives [4-6]. n-decylamine acts mainly as a cathodic inhibitor for the corrosion of zinc in acidic solutions

[7], in addition; its action was more understood on the basis of its property as a surface active agent in relation to its chain length [8].

In the present work, the efficiency of n-decylamine silicate:$CoCl_2$ and Bis-(diphenylthiocarbazono)mercury(II) complex as biocides to epoxy paints were tested, results of which open new fields of the much acclaimed and needed improvements towards available biocides with more aggressive adding up thermal stability, mechanical properties and also towards chemical agent's attacks.

Materials and methodology

1- Chemicals

Commercial epoxy paint, consists of (diglycidyl ether of bisphenol-A as resin and cycloaliphatic polyamine/reactive polyamide as hardener), were supplied by Yasmo Misr Company, Egypt. All other employed chemicals were supplied by Aldrich. The solvents used, namely, acetone, ethyl alcohol, chloroform, carbon tetrachloride were supplied by El-Nasr pharmaceutical chemicals, Abu-Zaabal, Cairo and were purified, distilled and kept under their specific drying agents.

2- Paint coat preparation specimens

All devised epoxy paints were applied to mild steel sheet specimens. The steel specimens were cut from the same batch of AS 37 low carbon steel sheet stock with a chemical composition of C: 0.04%; Si: 0.01%; Mn: 0.17%; P: 0.002%; S: 0.005%; Cr: 0.04%; Mo: 0.03%; Ni: 1.31% and Fe: balance). The steel specimen was cut to size of 5cm x 7.5cm (area = 37.5 cm^2). Before paint application, the steel substrate specimens were degreased with acetone, and pickled using concentrated hydrochloric acid to remove any corrosion remains of iron oxides. All steel specimens were emery brazed to remove all residues and finally washed with ethyl alcohol then placed in the desiccators for drying and conditioning before epoxy coated was applied.

3- Synthesis of paint additives
- **n-Decylamine silicate (DAS)**

n-Decylamine silicate was prepared by mixing 20 parts by weight of fine granular hydrated silica, 30 parts by weight of amine, 2 parts by weight of sodium carbonate and remaining parts by weight of water and then refluxed at 100°C for 1h. The water was then evaporated, thereby producing amine silicate. The product obtained is a granular n-decylamine silicate [9] designated in Table 1 as A.

- **Bis-(diphenylthiocarbazono) mercury(II) complex (Hg (HDz)$_2$)**

A 100 ml aqueous solution containing 2.02 m.moles of mercuric chloride salt was extracted with a solution of 3.5 m.moles of dithizone dissolved in 200 ml of chloroform by shaking the aqueous solution with portions of the organic solution in a separating funnel at room temperature. The reaction usually took place readily and the complex extracted quantitatively into the organic phase. This organic phase was then separated, dried and evaporated to dryness. The residue was recrystallized from carbon tetrachloride and the pure product was finally dried producing a dark red granular compound [10] designated as B in Table 1.

The chemical structure of bis-(diphenylthiocarbazone) mercury(II) complex

Table 1
Additives and their colors

Symbol	Additive type	Color
A	n-decylamine silicate (DAS)	White
B	bis-(diphenylthiocarbazono)mercury(II) complex (Hg (HDz)$_2$)	Dark red

Table 2
The composition of paint samples in the presence of different additives

Sample	Paint composition	Color
I	Native epoxy paint (resin : hardener = 4:1)	white
II	Epoxy paint + 10% additive (1mole DAS :1mole CoCl$_2$)	Faint grey

III	Epoxy paint + 5% additive (Bis-diphenylthiocarbazono)mercury(II) complex) (Hg (HDz)$_2$)	Bright dark red

4- Preparation of paint samples
- **Native epoxy paint**

8 grams of epoxy resin was added to 10 ml of acetone and left on magnetic stirrer for 30 min. 2 grams of epoxy hardener was added to this solution and left on stirrer for 20 min. The paint sample was then ready to be applied and referred to as I in Table 2.

- **Epoxy paint added DAS:CoCl$_2$**

3.54 m. mole of CoCl$_2$ was dissolved in 5 ml acetone then added to solution of 3.54 m. mole of amine silicate in 5 ml acetone where this mixture is stirred on magnetic stirrer for 30 minutes. 8 gm of epoxy resin was added to the above mixture and was left on stirrer until it became a homogeneous mixture (about 30 minutes). Then it was added to 2 gm of epoxy hardener and was left on stirrer for 20 minutes. This mix occurred with different ratios of additives to the epoxy paint according to that given in Table 2. After these steps the paint samples were ready to be applied. This paint composition sample is referred to as II in Table 2.

- **Epoxy paint added Hg (HDz)$_2$**

A 10 ml acetone containing 0.5 g of Hg (HDz)$_2$ complex was left on magnetic stirrer for 10 min. 8 grams of epoxy resin was added to this solution and stirred for 30 min. The mixture was then added to 2 g of epoxy hardener and stirred for 20 min. upon which the paint sample was ready to be applied. This paint composition sample is referred to as III in Table 2.

It should be noted that the concentrations of additives were the optimum ones chosen from a series of different concentrations list.

5- Structural conformation

All materials used of epoxy resin hardener, different additives and the hardened paint were subjected to FT-IR (Fourier Transform-Infra Red Spectroscopy) for structural conformation using Nicolet 6700 Thermo Scientific FT-IR.

6- Physico-mechanical properties

Free film mechanical properties of the coatings, such as bending, adhesion and impact were conducted according to ASTM procedures [1, 11] at Petroleum Pipes Co., Moustorood, Cairo, Egypt.

7- Thermogravimetric analysis (TGA)

The weight loss of all epoxy paint samples (native and modified) due to the effect of heat was monitored using Thermal Gravimetric Analyzer (TGA) Shimadzu-50, atmosphere N_2 at temperature rate of 5 °C/min.

8- Chemical resistance

The chemical resistance tests of all paint samples (native and modified) were performed according to ASTM procedures [6], in different mediums such as acidic medium (10 wt% HCl), alkaline medium (5 wt% NaOH) and artificial sea water (2.453 wt% or 24.53 g/l NaCl).

9- Photo Stability

(1×10^{-5})M of Hg (HDz)2 was prepared in pure chloroform. UV-Visible absorption spectrum of Hg $(HDz)_2$ solution was recorded firstly, and then the same solution is irradiated by different doses of light through different times of exposure. In each time of exposure, the UV-Visible absorption spectrum was recorded. The UV-Visible absorption spectra were recorded over the range (230–650 nm) using λ-Helios SP Pye-Unicam spectrophotometer. Continuous irradiation experiments were performed using a 150 W Xenon arc lamp (PTI-LPS-220 Photon Technology International) operated at 70 W.

UV-Visible absorption spectra were recorded over the range (200–1100 nm) using Photo dioide array (reflectance) IKS Polytec X-dap for the epoxy paint in presence of Hg $(HDz)_2$, sample V according to Table 2. Continuous irradiation experiments were performed using a 150 W Xenon arc lamp (PTI-LPS-220 Photon Technology International) operated at 70 W. The reflected beams are converted to absorbance beams through the following equation:

$$A = log10\ (1/R)$$

where, *A: absorbance;*   *R: reflectance.*

10- Antimicrobial activity

The anti-microbial activity of all materials used including epoxy resin, its hardener, different additives and the hardened paint samples were tested against different Gram positive and Gram negative bacteria at the Microbiology Department, Faculty of Science, Ain-Shams University, Cairo.

i- **Bacterial clinical isolates**

A total of 12 clinical isolates including, 4 *Staphylococcus aureus* (Sa1-4), 4 *Pseudomonas aeruginosa* (Ps1-4), 2 *Escherichia coli* (Ec1 and 2), and 2 *Klebsiella pneumoniae* (K1 and 2) were kindly supplied by Dr. M. Ghobashy, Microbiology department, Faculty of Science, Ain Shams University.

ii- **Growth and maintenance of bacterial clinical isolates**

All bacterial clinical isolates were grown routinely on nutrient agar or in nutrient broth media. The isolates were routinely incubated at 37°C. Liquid cultures were grown in 250 ml conical flasks containing 100 ml of nutrient broth medium and incubated in an orbital shaker (120 rpm) at 37°C. For long term storage, bacterial isolates were grown to exponential phase and spin down at 13,000 rpm for 5 min. The pellet was re-suspended in 10% (v/v) sterile glycerol in sterile screw cap vials (Technical Service Consultants Ltd., UK) and stored at -20°C.

iii- **Preparation of bacterial inoculum**

A twenty four hours nutrient broth culture of tested microorganisms was grown in an orbital shaking incubator, centrifuged, washed twice with PBS and then standardized to approximately 10^6 CFU ml^{-1} using broth medium, during the assay of paint and antibiotic tests.

iv- **Preparation of paint discs**

After mixing paint ingredients thoroughly, drops of paints (50 µl) were loaded on clean plastic sheets in sterile area. The diameter of the discs ranged between 6-7 mm. Paint only without additives was prepared as control.

v- **Bacterial sensitivity test**

Standard disc agar diffusion method was carried out to determine both antibiotic resistant profiles for Gram positive and negative bacteria and detect the activity of paints as well as their constituents against the clinical bacterial isolates according to Cheesbrough [13]. Standard antibiotic discs (Methicillin, Vancomycin, Amoxicilline, Azithromycin and Sulperazon for Gram positive and Gentamycin, Chloramphenicol, Tobramycin, Amikin, Cefobid, Fortam, Tetracyclin, Augmentin, Unasyin, and Ampicillin for Gram negative) were placed uniformly on the surface of nutrient agar plates each seeded with 100 µl of 24 h bacterial culture prepared as mentioned herein. Plates were incubated at

37°C for 24 h. Diameter of inhibition zone represents sensitivity of the organism [14].

Paint discs were loaded on plates containing nutrient agar medium seeded with the 24 h tested organism (100 µl). From each constituent, 50µl was loaded onto sterile Whatman filter paper disks (9mm diameter). The solvents were allowed to evaporate from discs before testing to eliminate toxicity. After solvent evaporation, discs were placed onto nutrient agar plates previously seeded with the tested clinical isolate. Solvents (50 µl/disc) were used as a negative control. Plates were incubated at 37°C for 24 h and inhibition zones were detected by a clear zone around the disks.

vi- Statistical analysis

All microbiological statistical analyses in this study were carried out using Microsoft Excel 2007, Analysis Toolpack (Microsoft Corporation). All data were calculated from at least 3 replicates and the standard errors for each datum were plotted on the graph.

Results and discussion

1- FT-IR structural studies of epoxy paint and additives

- **Epoxy paint and DAS:$CoCl_2$ additive**

Figure 1 represents the FT-IR structure of epoxy resin (a), the polyamine hardener (b) while (c) illustrates the native epoxy paint after curing. The FT-IR profile shows clearly the characteristic bands of the epoxy paint which are 3394.3 cm^{-1} for the polymeric–OH and 2513.2 cm^{-1} for (–NH) groups that are basically distinguished as characteristic for the epoxy paint after the hardening. These results match those given by Kumar et al [12] for the native epoxy paint without additives. Figure 2 illustrates the FT-IR spectra of the DAS as free (a) and the epoxy paint in presence of (DAS:$CoCl_2$) (b). As observed, disappearance of some peaks has occurred due to the addition of the DAS:$CoCl_2$ indicating a direct chemical reaction between the n-decylamine and the epoxy paint probably according schemes 1 and 2.

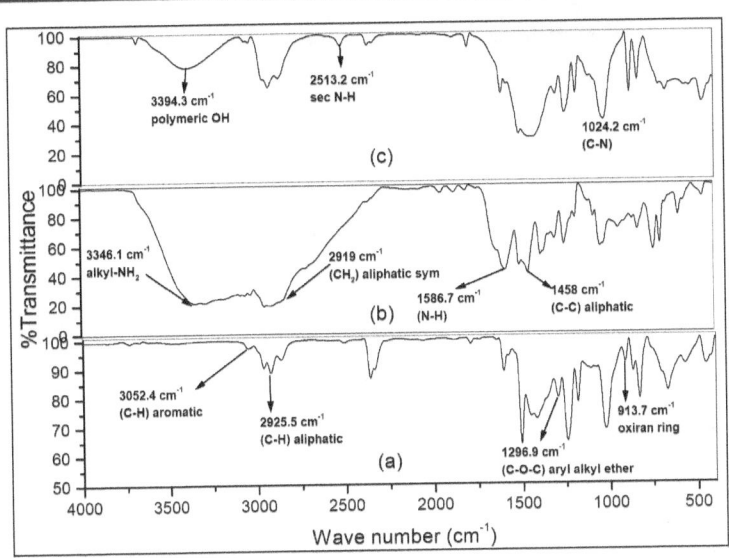

Figure 1
FT-IR spectra of (a) epoxy resin, (b) hardener and (c) native epoxy paint after curing.

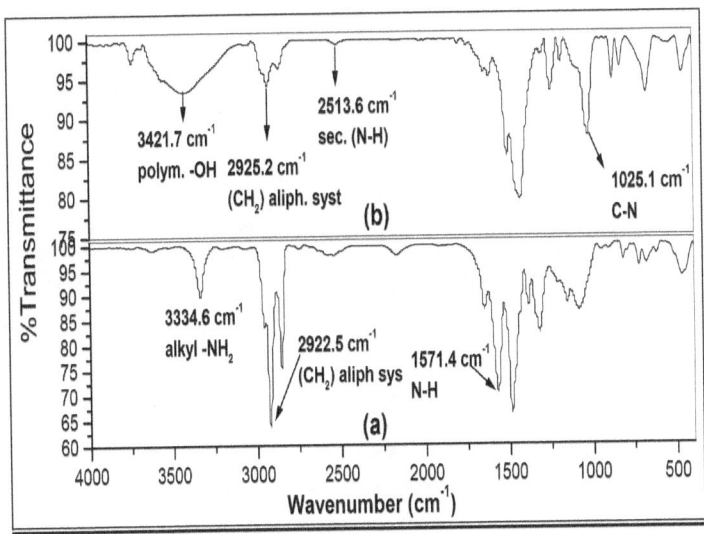

Figure 2
FT-IR spectra of (a) DAS, (b) epoxy paint in presence of DAS:$CoCl_2$.

Scheme 1 shows the direct reaction of the n-decylamine group with the epoxide ring of the epoxy resin which could in turn lead to ring opening that could contribute to increased crosslinking during the curing step (presence of polyamine hardener) [15]. This proposal could be demonstrated as follow:

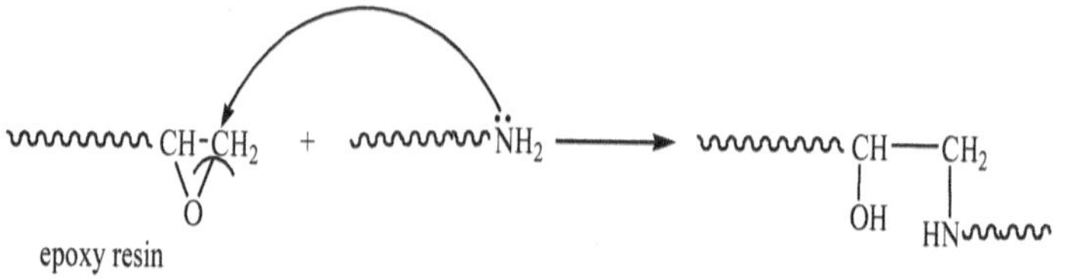

Scheme 1

The second scheme is the reaction of the n-decylamine with the epoxy resin during the curing step producing H_2O as a by-product as represented herein:

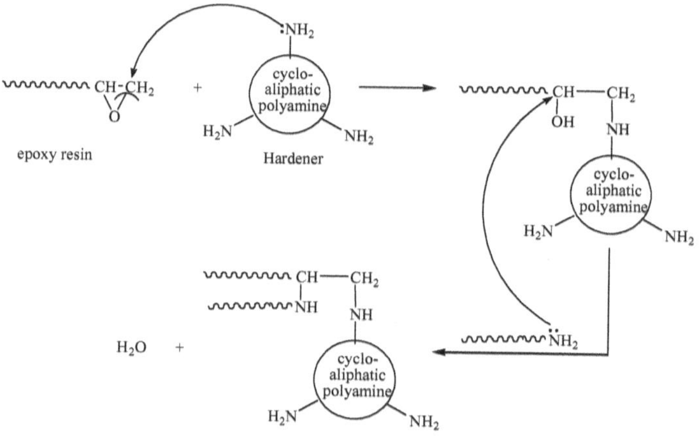

Scheme 2

The two proposed schemes could occur simultaneously or alternately according to availability of reactive sites evidenced by the –OH polymeric band shift from 3394.3 cm^{-1} to 3421.7 cm^{-1}, Fig. 2.

- **Epoxy paint in presence of Hg (HDz)$_2$**

Figure 3 illustrates the FT-IR spectral absorption chart for (a) Hg (HDz)$_2$ and (b) epoxy paint in presence of Hg (HDz)$_2$. The –NH (stretching) at 3230.3 cm^{-1} observed as a main structural feature of the Hg (HDz)$_2$, Fig. 3b, is agglomerated in polymeric –OH peak of position 3426 cm^{-1} in presence of the epoxy paint. In addition, it must be critically observed that the characteristic peaks representing –NH (bending), N-C-S and N-ph have agglomerated into a broader peak at 1431.7 cm^{-1} that corresponds to the C–C (aliphatic) indicating a preferential reaction during the curing step forming a cured epoxy resin of the form through HN-ph to –S.

$$\begin{array}{cc} -CH_2 & -CH- \\ | & | \\ -NH & OH \end{array}$$

This is indicated by the color consistency of the mercury complex compound and absence of any new peaks except for the broadening of a peak at the range 1300-1595 cm^{-1} and the shift occurred for polymeric –OH peak from 3394 cm^{-1} to 3426 cm^{-1}. The weakness of H-bond (S....H–N) in activated form (i.e., in solution) was demonstrated by Meriwether et al [10].

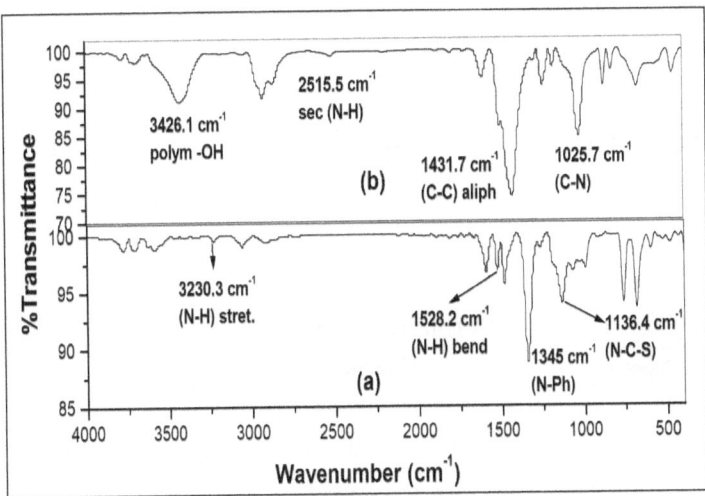

Figure 3

FT-IR spectra of (a) Hg (HDz)$_2$, (b) epoxy paint in presence of Hg (HDz)$_2$.

The intervention of the Hg (HDz)$_2$ complex with the epoxy paint product could be visualized mechanistically as illustrated in scheme 3.

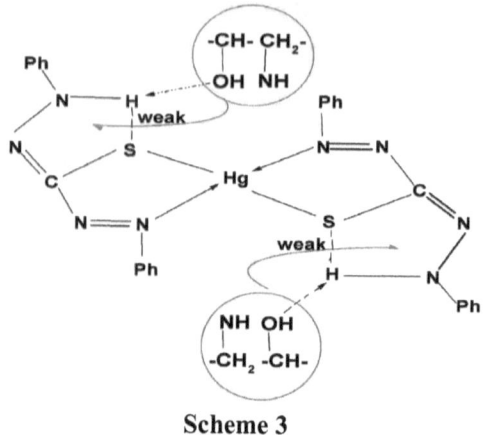

Scheme 3

2- Physico-mechanical properties

Free film mechanical properties of the coatings, such as bending, adhesion and impact were determined using the coated steel panels, Table, 3. All samples were examined according to ASTM D 3281-04 for bending test, (IS: 101 part5/sec.3, 1988) for impact resistance and ASTM C 1624 -05 for adhesion test. The results of the mechanical properties for painted samples are tabulated in Table 3.

Table 3

The mechanical properties of epoxy paint samples.

Sample designation	Impact at 25 cm	Impact at 90 cm	Adhesion	Bending
I	Excellent	Excellent	Excellent	Excellent
II	Excellent	Good	Excellent	Excellent
III	Excellent	Excellent	Excellent	Excellent

The above results indicate that even when different additives are used the modified epoxy paint still preserve its major mechanical properties, compared to the native epoxy paint, except for the impact test at height of 90 cm of paint sample II which shows slight value decline. Such decline could be explained as due to excessive crosslinking and the probable lack of elasticity of the modified epoxy paint as a probable direct effect of DAS:CoCl$_2$ and the special rubbery function of Si containing materials leading to compensating the lack of elasticity caused by addition of DAS:CoCl$_2$. This indication agreed with Kumar et al [12], who found that the reaction of the epoxy resin with γ-aminopropyltriethoxysilane to produce siloxane prepolymer led to modification of poor impact strength of epoxy resin itself. However in our case, the impact strength is slightly decreased in presence of Si containing materials due to presence of DAS which lead to excessive crosslinking.

3- Thermogravimetric analysis (TGA)

The results of the weight loss of all paint samples due to the effect of heat were monitored as exhibited in Fig. 4 and Table 4. Accordingly, the following conclusions are drawn:

- Native epoxy paint sample illustrates 3 stages of weight loss by temperature increase, Fig. 4a. In the first stage, degradation starts at 103°C with about 4% weight loss which could most probably indicate loss of incorporated water molecules that might be defined as by-product due to condensation. The second stage starts at a high temperature of 259°C with a weight loss of about 10 % with its final stage starting at 288 °C at a weight loss of 12.34 % with an average rate of degradation of 2.89 weight loss % /min. Such outstanding behavior is considered adequate for most applications.

- For modified epoxy paint with DAS: CoCl$_2$, (Fig. 4b) degradation effect starts at a much higher temperature of 185°C (a retention period with an outstanding starts complete curing and absence of any incorporated water which could reflect on a tighter microstructure of the paint) experiencing three stages of degradation at an average rate of 1.89 weight loss %/min. Such results indicate an increased thermal stability of the modified paint due to an acclaimed increased crosslinking degree according to either scheme 1 or 2 and presence of Si containing materials. The results encourage then the use of the DAS:CoCl$_2$ additive to the epoxy paints especially when surface applications are needed for a sustaining paint as such for thermal vigorous applications where the delayed degradation up to 185°C is sometimes required. These results are also demonstrated by Ahmed et al [1] who reached that using of siloxane in epoxy paints leads to noticeably enhance the thermal stability of epoxy paint. Epoxy novolac resin possesses good heat resistant property as compared to conventional epoxy resins. A thermally stable UV-curable epoxy coating has been developed by using functional amino silanes, viz., N-[3-(trimethoxysily1)propyl]ethylenediamine and a copolymer of aminopropylmethyl dimethyl siloxane in definite ratio. It has been observed that the use of silicon compounds invariably enhances the thermal stability [16].

- The modified epoxy paint with Hg $(HDz)_2$ illustrates a different thermogram where it starts to degrade at a mere similar temperature to that evidenced by the native epoxy paint first stage at 107°C. Yet, the second stage of degradation experienced less resistance to heat compared to the samples I and II. Such early degradation could be elucidated as due to the formation of a less stable cured product than that suggested when using DAS:$CoCl_2$ which helps concluding the formation of a product of less degree of crosslinking suggesting the intervention of mercury in decreasing crosslinking of the epoxy chains. In addition, modified epoxy paint with Hg $(HDz)_2$ witnessed the least average rate of weight loss of 0.79 %/min. which could be explained on the bases of mercury compound stability itself at that temperature range. As discussed above, the Si containing materials have the strong effective role in increasing the thermal stability of epoxy paint [1,16].

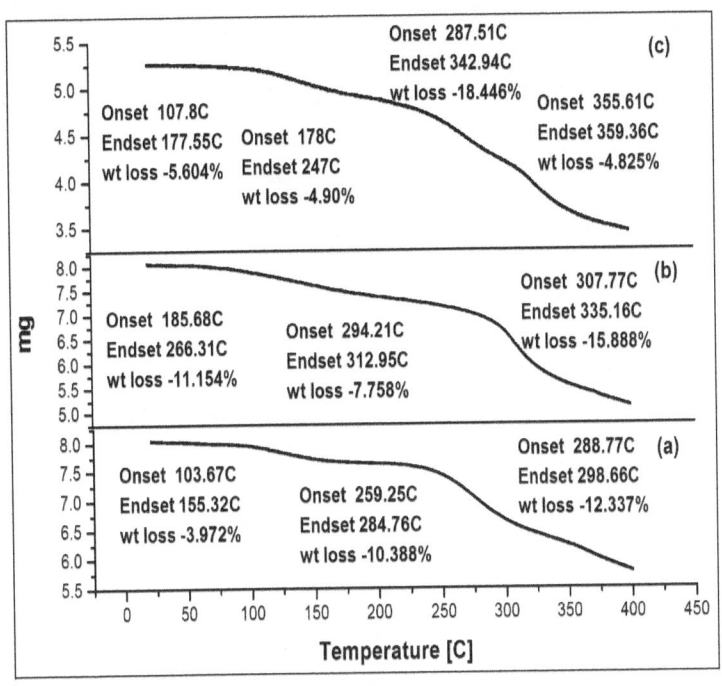

Figure 4

Thermo-gravimetric analysis of (a) native epoxy paint, (b) epoxy paint in presence of DAS:$CoCl_2$ and (c) epoxy paint in presence of ($Hg(HDz)_2$).

Table 4

Thermogravimetric studies (TGA) of epoxy paint samples.

Stages of weight loss		I Blank (native epoxy paint)	II (Epoxy paint + DAS:CoCl$_2$)	III (Epoxy paint + (Hg (HDz)$_2$)
First Stage	Temperature range of weight loss	103 – 155 °C	185 – 266 °C	107 - 177 °C
	Weight loss %	3.972	11.154	5.604
	Rate of weight loss % / minute	0.39	0.69	0.49
Second Stage	Temperature range of weight loss	259 - 284 °C	294 - 312 °C	178 - 247 °C
	Weight loss %	10.39	7.76	4.90
	Rate of weight loss % / minute	2.04	2.07	0.36
Third Stage	Temperature range of weight loss	288 - 298 °C	307 - 335 °C	287 - 342 °C
	Weight loss %	12.34	15.89	18.45
	Rate of weight loss % / minute	6.24	2.90	1.66
Fourth Stage	Temperature range of weight loss	-	-	355 - 359 °C
	Weight loss %	-	-	4.825
	Rate of weight loss % / minute	-	-	0.64
Average rate of thermal degradation % / minute		2.89	1.89	0.79

In general, addition of DAS:CoCl$_2$ or Hg (HDz)$_2$ as additives to epoxy paints imparts variable thermal profiles that could be functional according to required applications, with regard to either the start range of the first stage of degradation or supporting heat as observed when adding Hg (HDz)$_2$.

4- Chemical resistance

The chemical resistance tests of all paint samples; native and modified; were performed in acidic medium (10 wt% HCl), alkaline medium (5 wt% NaOH) and artificial sea water (2.453 wt% or 24.53 gm/l NaCl). It must be noted that the data of the native epoxy paint employed is reported as excellent resistance towards water, alkali, inorganic acids while poor against ketones and glycol ethers.

5- Resistance to alkali and acid attack

The results of testing native and modified epoxy paint samples in alkali and acid solutions are summarized in Tables 5, 6 and Fig. 5 and 6. It is concluded from the results that sample III (modified epoxy paint by Hg $(HDz)_2$) is quite resistant to alkali attack in a profile similar to that of native epoxy paint while that modified by DAS:$CoCl_2$ was less efficient in resisting the alkali effect in the sense of weight loss which could be simply explained as due to direct reaction of the paint with $CoCl_2$ or SiO_2 species forming soluble products in the alkaline medium. The alkali resistance of epoxy paint modified by Hg $(HDz)_2$ was explained probably due to stability of Hg $(HDz)_2$ itself toward alkalis.

The modified epoxy paint by DAS:$CoCl_2$ seems to exhibit better resistance towards HCl acid (10%) than native epoxy paint sample due to most probably the presence of SiO_2 which resist the effect of acids in general. The results indicate that all native and modified epoxy paint samples decreased to nearly their third original weight after duration of 70 days immersion. Nevertheless, some of them have reached this loss in weight in less time (blank and sample III). The progress of acid attack in presence of DAS:$CoCl_2$ seems slower than those of the others due to the role of SiO_2. The addition of Hg $(HDz)_2$ to epoxy paint, also led to increase the stability of epoxy paint toward acids attack in a profile lower than that in presence of DAS:$CoCl_2$ due to stability of Hg $(HDz)_2$ itself toward acids. Epoxy paint in presence of polyamide are characterized by lower alkali resistance than that siloxane-modified-epoxy paint in presence of polyamide where the later was found unaffected by NaOH solution even after 7 days of immersion. This indication was due to the intervention of (Si–O) into the essential backbone of network polymer [1]. On the contrary, in our present work and due to the effect of SiO_2 is attachment to n-decylamine and not in backbone of network polymer, the compounds formed between the SiO_2 and epoxy paint were observed soluble in alkaline medium.

The curing products of epoxy paint with *p*-phenlenediamine or pentaethylenehexamine as curing agents were found to be sensitive to both aqueous acidic and alkaline solutions [4]. With *p*-phenlenediamine, the data indicate the high resistance possessed by it towards both acidic and alkaline solutions. In the case of pentaethylenehexamine, although the cured products have the same groups (imine and ether groups), yet it was found to be less sensitive to this kind of attacks. In the former case, the resistance was referred to higher crosslink density of networks due to high epoxy functionalities [4]. This observation can also apply in our work due to excessive crosslinking of epoxy paint in presence of DAS:$CoCl_2$ in addition to the role of SiO_2.

Table 5

Effect of 5 % NaOH on epoxy paint samples by weight loss technique

Sample	W L % after 10 days	W L % after 20 days	W L % after 30 days	W L % after 40 days	W L % after 50 days	W L % after 60 days	W L % after 70 days
I Blank epoxy paint	4.89	5.89	6.12	6.36	6.47	6.52	6.62
II (Epoxy paint + DAS:$CoCl_2$)	W L % relative to blank*						
	After 10 days	After 20 days	After 30 days	After 40 days	After 50 days	After 60 days	After 70 days
	1.16	1.52	1.93	2.28	2.71	3.11	3.37
III (Epoxy paint + Hg(HDz)$_2$)	0.74	0.79	0.86	0.94	0.95	0.95	0.94

W L has been calculated as the weight difference percent of the sample after been subjected to the alkali effect duration.
* Relative to unity.

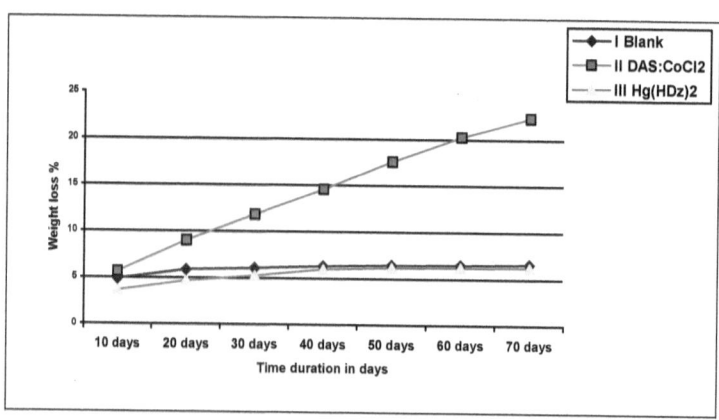

Figure 5

Relationship between the weight loss % of immersed paint samples in (5% NaOH) and immersion time for alkali resistance.

Table 6

Effect of 10% HCl acid on epoxy paint samples by weight loss technique

Sample	W L % after 10 days	W L % after 20 Days	W L % after 30 days	W L % after 40 days	W L % after 50 days	W L % after 60 days	W L % after 70 days
I Blank (epoxy paint)	18.76	28.5	36.02	37.83	37.90	38.06	38.09
II (Epoxy paint + DAS:CoCl$_2$)	W L % relative to the blank*						
	After 10 days	After 20 days	After 30 days	After 40 days	After 50 days	After 60 days	After 70 days
	0.747	0.628	0.665	0.706	0.813	0.898	0.962
III (Epoxy paint + Hg (HDz)$_2$)	1.24	1.15	0.929	0.896	0.897	0.896	0.913

W L has been calculated as the weight difference percent of the sample after been subjected to the acid effect duration periods.
 * *Relative to unity.*

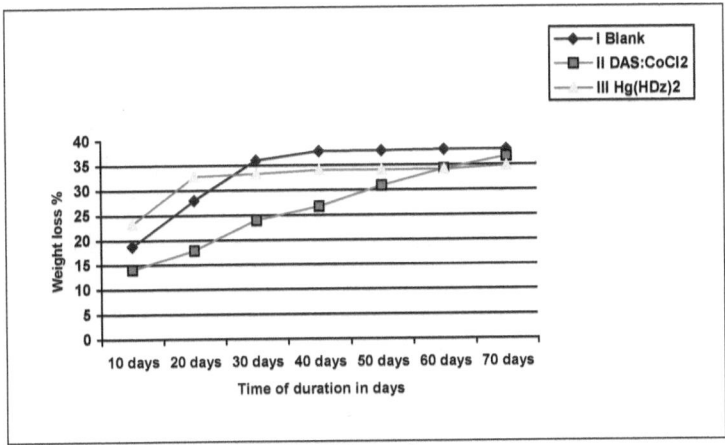

Figure 6
Relationship between the weight loss % of immersed paint samples
in (10% HCl) and immersion time for acid resistance.

6- Resistance to artificial sea water

The results of testing native and modified epoxy paint samples in artificial sea water are summarized in Table 7 and Fig. 7. As surprisingly observed from the results given, native epoxy paint shows the best sustaining results at very low weight loss ratios even after 70 days immersion of 5.66 %.

Table 7

Effect of artificial sea water on epoxy paint samples by weight loss technique.

Sample	W L % after 10 days	WL % after 20 Days	W L % after 30 Days	W L % after 40 days	W L % after 50 days	W L % after 60 days	W L % after 70 days
I Blank (epoxy paint)	3.86	4.53	4.89	5.1	5.21	5.22	5.66
II (Epoxy paint + DAS:CoCl$_2$)	W L % relative to blank*						
	After 10 days	After 20 days	After 30 days	After 40 days	After 50 days	After 60 days	After 70 days
	1.35	1.66	1.69	1.88	1.84	1.94	1.95
III (Epoxy paint + Hg (HDz)$_2$)	1.26	1.37	1.40	1.36	1.33	1.33	1.23

W L has been calculated as the weight difference percent of the sample after been subjected to the artificial sea water effect duration.
* *Relative to unity.*

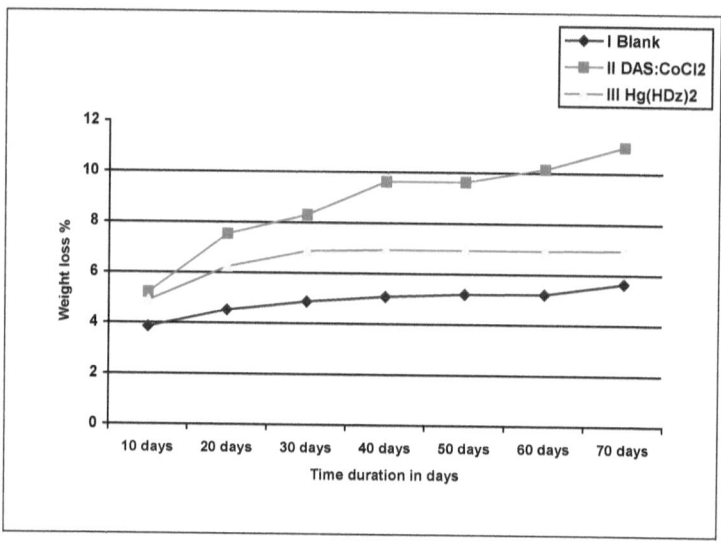

Figure 7
Relationship between the weight loss% of immersed paint samples and immersion time for artificial sea water resistance.

Addition of DAS:CoCl$_2$ to the epoxy paint enhanced the attacking power of the sea water corrosive effect due to a possible pH medium facilitation for alkaline earth reaction through the presence of SiO$_2$ and probably of faster reaction than that shown by its continuous increases of weight loss to nearly double that shown by the native epoxy paint. The addition of Hg (HDz)$_2$ to the epoxy paint led to increasing the weight loss than that of the native epoxy paint which increases possibly due to the presence of aggressive anion, Cl$^-$ effect that might enhance the solubility of the species. Nevertheless, the modified epoxy paint with Hg (HDz)$_2$ is found more stable toward artificial sea water than that modified by of DAS:CoCl$_2$. Hamdy [17] found that the epoxy primer fluoropolymer top coat specimens showed a dramatic increase in the corrosion rates under scratched conditions after less than 30 days of immersion in NaCl but the corrosion resistance was improved upon surface pre-treatment with vanadate due to formation of a highly protective oxide layer which improve the corrosion resistance by immersion in artificial sea water. Our results indicate that the absence of any protected layers on the surface of our paint samples and presence of Cl$^-$ of sea water led to decreasing the artificial sea water resistance [17]. Figure 7 represents the pattern of artificial sea water attack to various epoxy paint samples.

7- Photo Stability

It is also of interest to study the photo stability of the used mercury complex. Figure 8 illustrates the photo stability of the mercury complex. The mercury complex solution is stable toward continuous irradiation to the Xenon arc lamp (7000 wat/m). Until 20 min of irradiation, this stability deteriorated and the complex photodegradation may be due to the breaking of the essential back bone of the complex upon irradiation which is evidenced by changing the color of the complex solution without restoring the initial color. This is not the case when the benzene solution of the complex is just irradiated by sunlight (930 wat/m) due to the small flux.

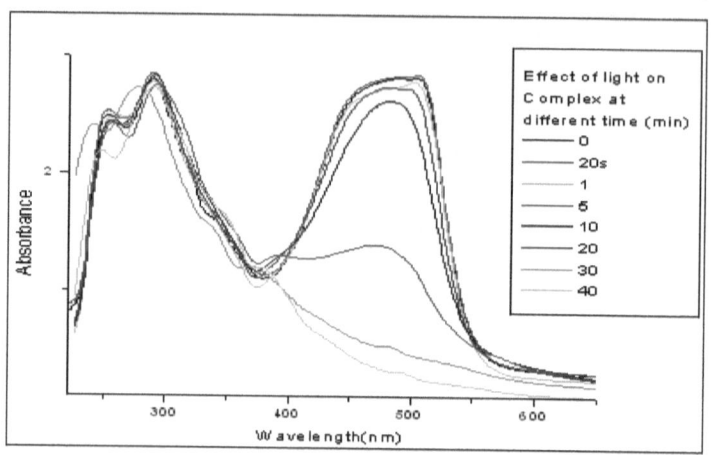

Figure 8
The photo stability of Hg (HD$_z$)$_2$ complex solution.

Figure 9 illustrates the photo stability of the modified epoxy paint in presence of Hg (HDz)$_2$. The addition of epoxy paint to the mercury complex slightly increased the stability of Hg (HDz)$_2$ embedded in the paint itself towards photodegradation which could be due to the formation of H-bond between the cured structure of the epoxy resin and the Hg (HDz)$_2$. Such bonding could lead to weakening of the other H-bond of (S…H), Scheme 3. Such modification of H-bonding could be considered as responsible to the evidenced prolonged stability. An early claim [18] of NH–ph resonance form due mainly to the S….H was considered responsible for such photo resistance of Hg (HDz)$_2$ itself.

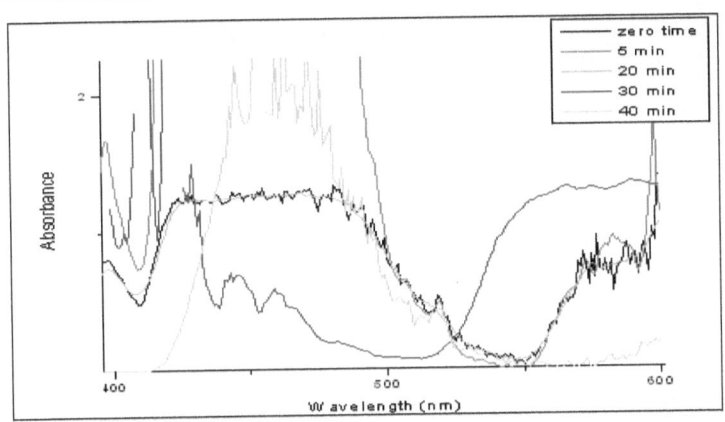

Figure 9
The photo stability of epoxy paint modified by Hg $(HD_z)_2$.

8- Anti-microbial activity examination

Sensitivity of the supplied clinical isolates, Table 9, towards different standard antibiotics was carried out to confirm the multi-drug resistance of these isolates. All isolates were multi-drug resistant; in addition, all *Staphylococcus aureus* isolates were vancomycin and (VRSA) methicillin (MRSA) resistant isolates.

The antimicrobial activities of paints were carried out by disc agar diffusion method. Paints containing either $DAS:CoCl_2$ or Hg $(HDz)_2$ showed activity against Gram positive and Gram negative clinical isolates with different extents, while incorporating other chemicals into epoxy paint did not show significant activities against all tested isolates. All *Staphylococcus aureus* isolates (Sa1-4) were significantly inhibited by both types of paints with a diameter of inhibition zones ranging from 1.3 ± 0.05 to 1.88 ± 0.15 cm for paint sample III and 1.4 ± 0.15 to 2.35 ± 0.2 for paint sample II. The lowest activity was exhibited by strain Sa4 for sample III and Sa1 for sample II respectively (Table 8, Fig. 10, 11).

For *Pseudomonas aeruginosa* isolates, 2 out of 4 isolates were significantly inhibited by both types of paints with relatively the same degree. In addition, the inhibitory capability of the 2 paint samples against *Pseudomonas aeruginosa* isolates was less than *Staphylococcus aureus* isolates (Table 8, Fig. 10 and 13). On the other hand both paints did not show any activity against all

Escherichia coli and *Klebsiella pneumoniae* clinical isolates. All separate ingredients showed high inhibitory action against all tested isolates, however on mixing with each other the activity was reduced or demolished. Testing the paint samples against the same strains after 1, 2, 3, 4, 8 and 12 weeks of preparation showed no reduction in their activities.

Table 8

Diameter of inhibition zones of different antibiotics (cm) against 12 clinical isolates.

Bacterial isolates	GN 10 µg	C 30 µg	TOB 10 µg	Ak 30 µg	CFP 30 µg	CAZ 30 µg	TE 30 µg	AMC 30 µg	SAM 20 µg	AMP 30 µg	V	Me	AML	AZM	SCF
Sa1	ND	ND	ND	ND	ND	ND	ND	ND	ND	ND	0	0	0	0	0
Sa2	ND	ND	ND	ND	ND	ND	ND	ND	ND	ND	0	0	0	0	0
Sa3	ND	ND	ND	ND	ND	ND	ND	ND	ND	ND	0	0	0	0	0
Sa4	ND	ND	ND	ND	ND	ND	ND	ND	ND	ND	0	0	0	0	0
Ps 1	0	0	0	0	1.6	0	0	1.1	1.2	0	ND	ND	ND	ND	ND
Ps 2	0	0	0	0	0	0	0	0.9	0	0	ND	ND	ND	ND	ND
Ps 3	0	0	1.6	0	0	0	0	1.2	0	0	ND	ND	ND	ND	ND
Ps 4	0	0	0	1.7	1.1	0	0	0.9	1.3	0	ND	ND	ND	ND	ND
E 1	1.7	0.9	0	2.3	0	0	0	1.2	0	0	ND	ND	ND	ND	ND
E 2	0.9	0	0	0	0	0	0	0.9	0	0	ND	ND	ND	ND	ND
K 1	0	0	0	1.8	0	0	0	1.1	0	0	ND	ND	ND	ND	ND
K 2	0	0	0	1.6	0	0	0	1.3	0	0	ND	ND	ND	ND	ND

GN: Gentamycin, C: Chloramphenicol, TOB: Tobramycin, Ak: Amikin, , CFP: Cefobid, CAZ: Fortam, TE: Tetracyclin, AMC: Augmentin, SAM: Unasyin, AMP: Ampicillin, V: Vancomycin, Me: Methicillin, AML: Amoxicilline, AZM: Azithromycin and SCF: Sulperazon and ND: not done.

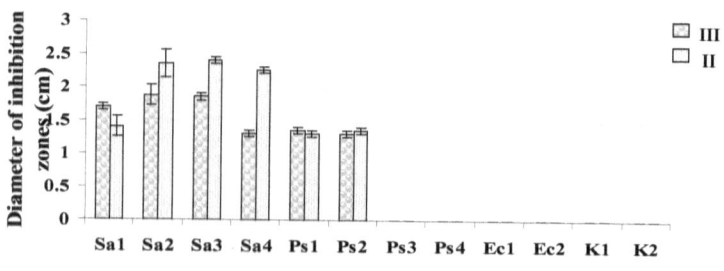

Figure 10

Antimicrobial activities of paint samples against 12 clinical isolates.
III, paint disc impregnated with Hg (HDz)$_2$, II, paint disc impregnated with DAS:CoCl$_2$

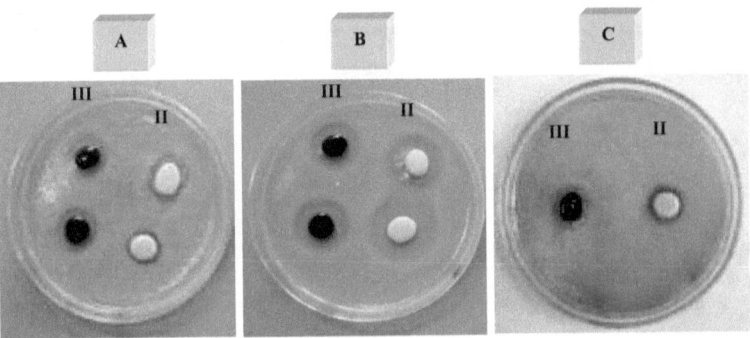

Figure 11

Antimicrobial activities of paints against A) *Staphylococcus aureus* Sa1, B) *Staphylococcus aureus* Sa2 and C) *Pseudomonas aeruginosa* Pa1. III, paint disc impregnated with Hg (HDz)$_2$, II, paint disc impregnated with DAS:CoCl$_2$.

DAS:CoCl$_2$ and Hg (HDz)$_2$ as additives to epoxy paints were investigated for the purpose of imparting rich physico-mechanical and chemical properties to an epoxy paint. The addition of DAS:CoCl$_2$ or Hg (HDz)$_2$ to epoxy paints demonstrated specific thermal profiles that is diverted from the native paint and could be utilized at preference of temperatures of application. Addition of bis-(diphenylthiocarbazone)mercury(II) complex was found advantageous when suffering alkali attack in a profile stronger than that of native epoxy paint and that modified by the n-decyalmine silicate:CoCl$_2$. Both modified by bis-(diphenylthiocarbazone)mercury(II) complex and n-decyalmine silicate:CoCl$_2$ were found resistant to acids compared to the native epoxy paint. Bis-(diphenylthiocarbazone)mercury(II) complex and n-decyalmine silicate:CoCl$_2$ additives to the epoxy paint were found less resistant to artificial sea water than that of native epoxy paint. Upon sunlight irradiation of Hg (HDz)$_2$ complex solution, the color of the complex changes from the orange-yellow of the normal form to the blue color of the activated form which could impart excessive activity as anti-microorganisms. The addition of epoxy paint to the Hg (HDz)$_2$ led to increasing the photo stability of complex solution. Such remark brings the attention for improving such ability especially and those additives proved excellent ability as antimicrobial agents (anti-bacteria) and strengthening such ability will favor the use of modified paints by bis-(diphenylthiocarbazone) mercury(II) complex and n-decylamine silicate:$CoCl_2$ as excellent applicator to protect steel works in marine and medical activities.

Microbiological argument

Staphylococcus aureus has long been recognized as an important human pathogen causing both nosocomial and community-acquired infections (K3). *Staphylococcus aureus* (MRSA) and (VRSA) have emerged as an important cause of hospital-acquired infections worldwide (K1, K2). Other microorganisms including *Pseudomonas aeruginosa* are frequently life threatening and often challenging to treat. *Pseudomonas aeruginosa* is one of the leading gram-negative organisms associated with nosocomial infections. Hygiene coatings may be defined as a coating in which antimicrobial agent is incorporated to offer protection to surfaces which might be subjected to microbial growth.

The activity of 2 different paint samples to inhibit microbial growth was demonstrated against 12 clinical isolates isolated from patients suffering from serious illness. Both paints inhibited growth and colonization (biofilm formation) of both Gram positive (*Staphylococcus aureus*) and Gram negative (*Pseudomonas aeruginosa*) bacteria with different extents. All methicillin and vancomycin resistant isolates of *Staphylococcus aureus* were susceptible to both paint samples, however, paint II was more effective compared with paint III. Since, paint II did not contain toxic material compared with paint III, paint II was preferable to be used compared with paint III. In addition, the two paint samples were able to inhibit the growth and colonization of 2 multi-drug resistant strains of *Pseudomonas aeruginosa*. Sensitivity of *Pseudomonas aeruginosa* strains was lower than *Staphylococcus aureus*. Antagonistic effect was detected after mixing all the additives with the other ingredients of both paints. This showed that after paint casting, the active additives were chelated in the epoxy paint thus reduce their activity. The activity of paint samples did not change after 3 months of preparation with the same efficiency. The results agreed with [19] that aliphatic amines (ingredient of II) were more efficient against the Gram positive bacteria.

Quaternary ammonium compounds (QACs) of C16 hydrophobic tail length affected the outer membrane of Gram-negative bacteria more extensively than shorter chain compounds, possibly due to the C16 chain interact strongly with the fatty acid portion of lipid A. It was also reported that monoalkyl QACs bind by ionic and hydrophobic interaction to microbial membrane surfaces, with the cationic head group facing outwards and the hydrophobic tails inserted into the lipid bilayer causing rearrangement of the membrane and subsequent leakage of intracellular constituents. It was Reported [20] that a common feature of QACs is their ability to cause cell leakage and membrane damage, primarily due to their adsorption in large amounts to the bacterial membrane [21].

Surface-active agents, including non-ionic surfactants, are known to disrupt cell membranes because they dissolve in both extracellular fluid and the lipid membrane. This lowers the surface tension of the membrane, allowing water to flow into the cell and ultimately resulting in lysis and bactericidal action. The balance between the hydrophilic and lipophilic sections of the molecules is essential for these processes. The irritating properties of cationic surfactants have been one of the major limitations to a widespread use of this type of surfactants with known bactericidal activity, in personal care products. In personal hygiene as well as in the cosmetic industry, the association of a low antibacterial activity with emulsifying potential is desirable in order to clean and simultaneously cause the least disturbance in the normal skin flora and moisture balance. Therefore, the antimicrobial activity of the new surface-active glycosides was also evaluated, using the paper disk diffusion method [22].

Part II

Nano ZnO/amine composites antimicrobial additives to Acrylic emulsion paints

Introduction

Long-term antimicrobial activity can be imparted in many coating formulations through the incorporation of nanomaterials. Zinc oxide is commonly used in pharmaceutical products to prevent or treat topical or systemic diseases owing to its antimicrobial properties [23]. ZnO nanoparticles were shown to have a wide range of antibacterial activities against both Gram-positive and Gram-negative bacteria, including major foodborne pathogens like *Escherichia coli*, *Salmonella*, *Listeria monocytogenes*, and *Staphylococcus aureus* [24-27]. It is necessary to understand the mechanism of ZnO action against bacteria, but to date, the process underlying their antibacterial effect is not clear. However, early studies suggested that the primary cause of the antibacterial function might source from the disruption of cell membrane activity [28]. Another possibility could be the induction of intercellular reactive oxygen species, including hydrogen peroxide (H_2O_2), a strong oxidizing agent harmful to bacterial cells [26, 29]. It has also been reported that ZnO can be activated by UV and visible light to generate highly reactive oxygen species such as OH^-, H_2O_2, and O_2^{2-}. Although the antibacterial mechanism of ZnO nanoparticles is still unknown, the possibilities of membrane damage caused by direct or electrostatic interaction between ZnO and cell surfaces, cellular internalization of ZnO nanoparticles, and the production of active oxygen species such as H_2O_2 in cells due to metal oxides were proposed in earlier studies [30,31].The effectiveness of the nano ZnO in many antimicrobial applications as in food preservation [32], insecticide and synthetic and natural textiles were also identified and reported including improvements to coatings properties, thermal and mechanical properties [32-35]. Such implemented direction of the ZnO is based on its safety function to health and enviroment.

In the present work, nano ZnO composites of four types of amines namely; 2-amino-1-(4-nitrophenyl)-1,3-porpandiol (PD), 3-amino-1,2,4-Triazole (T), diphenylamine (DPA) and N,N-dimethylamine (DMA) were evaluated as biocide to emulsion acrylic paints to emphasis their function through a probable cause of active oxygen species occurrence responsible for the ZnO antimicrobial function, the results were assessed with regard to the composite employed and the amine effect on the nano ZnO particles agglomeration or dispersion in addition to biological assessment of each amine composite for effectiveness comparison and preference focusing on the most safest amine to human and environment. .

Experimental and methodology

2. Chemicals

a- Acrylic emulsion paints; both biocide added and non-biocide and free types of paints were supplied by Paints and Chemical Industries Company "Pachin", El Obour city, Cairo. Solvents employed during the present work such as ethyl alcohol and N,N-dimethyl formamide (DMF) were supplied by El-Nasr pharmaceutical chemicals, Abu-Zaabal after being subjected to distillation, purification and drying prior use. ZnO and all other chemicals and amines employed were of Aldrich product.

b- Nutrient agar medium (Difco), Nutrient broth medium (Difco) and Phosphate buffer saline (PBS) were all employed by the biological assessment tests as standard materials.

2- Preparation and Synthesis

a- Preparation of ZnO nanoparticles

ZnO nanoparticles, of an average size of 30-95 nm, were prepared by grinding using ball mill technique available at the Egyptian Petroleum Research Institute (EPRI), Nasr City, Cairo. Rather the ball mill technique is not adequate to produce uniform ZnO particles in the nano form of 100 nm or less, it was employed, however, to accommodate the major aspect of using a broader scope of the ZnO and to foresee the effect of the amines on the oxide dispersion factor.

b- Synthesis of nano ZnO composite

Exact weights of ZnO nanoparticle sample were slowly added to the respective weights of the different amines namely; 2-amino-1-(4-nitrophenyl)-1,3-porpandiol (PD), 3-amino-1,2,4-Triazole (T), diphenylamine (DPA) and N,N-dimethylamine (DMA) dissolved in DMF heated up to 100°C. The whole mixture was then stirred for 24 h at room temperature. The mixture was finally dried under vacuum to remove the DMF solvent. The dry product is denoted as the nano ZnO composite with each of the amines. The ZnO/amine composites were thoroughly washed with bidistilled water to free the composites from any excess amines. 10% and 20% by weight of the ZnO to the total weight of the composite were prepared for activity comparison [37].

To affirm function and selectivity as a biocide additive, a series of weight % additives (1%, 2% and 3%) to raw paint were prepared and biologically tested.

3- Paint and additives: structural, thermal and antimicrobial features and characterization

a- Chemical structure elucidation using FT-IR analysis

Acrylic paint, different amine additives (biocides) and the paint/biocide were subjected to FT-IR (Fourier Transform-Infra Red Spectroscopy) structural conformation using Nicolet 6700 Thermo Scientific FT-IR available at the Central Laboratory of the Faculty of Science, Ain Shams University, Cairo.

b- Thermo-gravimetric Analysis (TGA)

The weight loss of all paint samples due to the effect of heat (thermal stability) was monitored using Thermal Gravimetric Analyzer (TGA) Shimadzu-50, under N_2 atmosphere at temperature rate of 10 degrees/minute available at the Central Laboratory of the Egyptian Petroleum Research Institute, Nasr City, Cairo.

c- Transmission Electron Microscope

TEM of the synthesized nano composites and ZnO nanoparticles were carried out at TEM unit of the Faculty of Science, Ain-Shams University, Cairo using JOEL, JEM 1200 EX available at the Central Laboratory of the Faculty of Science, Ain Shams University, Cairo.

d- Antibacterial Activity

The antibacterial activities of paint with different additives were tested against 2 multidrug resistant clinical isolates, namely; *Pseudomonas aeruginosa* and *Staphylococcus aureus* with 6 strains; 3 strains of *Pseudomonas aeruginosa* and 3 strains of *Staphylococcus aureus*, which were supplied by Microbiology Department, Faculty of Science, Ain Shams University, Cairo, and according to the following procedure:

i- Preparation of bacterial inoculum

A twenty four hours nutrient broth culture of tested bacteria was grown in an orbital shaking incubator (120 rpm) at 37°C and standardized to approximately 10^6 CFU ml^{-1} using nutrient broth medium.

ii- Preparation of paint discs

After mixing paint ingredients with and without additives thoroughly, drops of paints (50 µl) were loaded on clean plastic sheets in sterile area. After drying the discs were kept in clean plastic bags and kept away from sunlight till tested. Discs with paint only (without additives) were prepared as control.

iii- Antibacterial activity of paint discs

Two different tests were carried out to test the antibacterial activities of paints.

- Agar diffusion method

A standard disc diffusion method was used to detect the activity of paints and their constituents against the clinical bacterial isolates according to Cheesbrough and Adonizio et al., [38,39]. Paint discs were loaded on plates containing nutrient agar medium seeded with 100 µl of the 24 h tested organism. Plates were incubated at 37°C for 24 h and inhibition zones were detected by a clear zone around the disks.

- Turbidity method

Sterile tubes containing 5 ml nutrient broth medium each was inoculated with 20 µl of the tested organism. Five paint discs were transferred to each tube in sterile condition. Tubes received no discs acted as microbial growth control. All tubes were incubated in an orbital shaking incubator (100 rpm) at 37°C for 24 h. After incubation period the turbidity was measured at OD 600 nm using λ-Helios SP Pye-Unicam spectrophotometer.

iv- Antimicrobial Data Statistical Analysis

All statistical analysis in this study was carried out using Microsoft Excel 2000, Analysis Toolpack (Microsoft Corporation). All data were calculated from at least 3 replicates and the standard errors for each datum were plotted on the graph.

Results and discussion

1- Chemical structural analysis (FT-IR)

The FT-IR structural analysis of acrylic paint, ZnO, PD, T, DPA and DMA were conducted for their chemical structural confirmation, results of which found matching referenced data [40-47], and are manifested in Table 9.

As well noticed from the survey given in Tables 9 and 10, the amine additives in their distinct formulations did not interact covalently with either the ZnO or with the polymeric acrylic paint chains with any new bands evolving. Therefore, the biological activity assessment results should not be refereed to any chemical changes of the additives onto both the ZnO particles and/or the polymeric based paint molecules, but should be directed towards the influence of physical affiliations imposed within the paint environment such as:

- direct functional cause of the ZnO,
- ZnO particle size role,
- medium acidity/basicity and aromaticity parameter
- biological cell/ additive interaction accessibility based on a proposed active oxygen species production as thought responsible of microorganism's disability.

Table 9

FT-IR absorption spectral bands (cm⁻¹) of ZnO, amine additives and ZnO composites characterization

ZnO	PD	ZnO/PD	Gp.	T	ZnO/T	Gp.	DPA	ZnO/DPA	Gp.	DMA	ZnO/DMA	Gp.
	3373	3373	-OH	3413	3414	Asym .NH	3380	3380	-NH	3453	3431	Sec – NH
	3306	3306	-NH2	3332	3331		3038	3036	Arom - C-H	1638	1626	Bend. Sec - NH
	3074	3074	Arom. -C-H	3214	3214	Sym. -NH	1491	1490	C=C			
Finger print (447 cm⁻¹)	2958	2959	Aliph. C-H	1641	1641	NH₂ Sciss		Broad at 435	ZnO finger print		447	ZnO finger print
	1518	1518	Arom. -NO₂	1596	1597	-C-N str.						
				1535	1535	-N-N						
		Shoulder at 45	ZnO Finger print	1214	1214	Sec. - NH						
				-	436	ZnO finger print						

Amines: PD = 2-amino-1-(4-nitrophenyl)-1, 3-porpandiole, T= 3-amino-1,2,4-Triazole, DMA= dimethyl amine, DPA= diphenylamine
Composites: ZnO$_{PD}$ = ZnO propandiol composite, ZnO$_{DMA}$ = ZnO dimethylamine composite, ZnO$_T$ = ZnO triazole composite, ZnO$_{DPA}$ = ZnO diphenylamine composite.
Sym. = Symmetric, Asym. = Asymmetric, Broad= Broadening, Bend. =Bending, Str. = stretching, Sec=secondary, Arom.= Aromatic, Aliph=aliphatic

The effect of the nano ZnO and its composites on the structural features of the acrylic paint is also shown in Table 10.

Table 10

FT-IR spectra analysis (cm^{-1}) of basic poly acrylic paint and its ZnO/amine added characterization

Native acrylic paint	ZnO/PD + Paint	ZnO/T + Paint	ZnO/DPA + Paint	ZnO/DMA + Paint	Group
3364	3372	3331	Screened at broad 3380	3366	-OH Str.
			3380	Screened at broad 3366	Sec. -NH
-	3310	Shoulder at 3400	-	-	-NH$_2$
2954 2873	2954 2873	2954 2873	2954 2873	2954 2873	Aliph. -CH
1733	1732	1733	1732	1733	-C=O
-	-	1639	-	-	-NH$_2$ sciss
-	-	1550	-	-	-N-N
-	1521	-	-	-	-NO$_2$
1167	1168	1167	1167	1167	-C-O ester
1020	1020	1020	1020	1020	Cellulose added
500-700	500-700	500-700	500-700	500-700	TiO$_2$

2- Thermal stability (TGA)

Figure 12, illustrates the TGA profiles of dried basic acrylic paint films and nano ZnO composite additive paints. As revealed, thermal degradation of the paint films occurs at one step only at around 350 °C for nearly all samples, at a mere equal weight loss of 34.9 – 37.4 %. The weight loss is mostly inferred to the decomposition of the binder material, acrylic polymer and the ratio of additives to the acrylic paint [40]. The slight differences of weight loss% when comparing the ZnO composite added paint films could probably due to the indulged agglomeration or dispersion caused by the amine presence and as noted from the TEM images, Fig. 13. The TGA results confirm the persistence of the thermogram of either the pure or ZnO/amine composites added paints excluding the possibility chemically modified material formation [36]. The above results indicate that the ZnO/amine composites of different natures had the slimmest effect on the basic acrylic emulsion paint thermal stability.

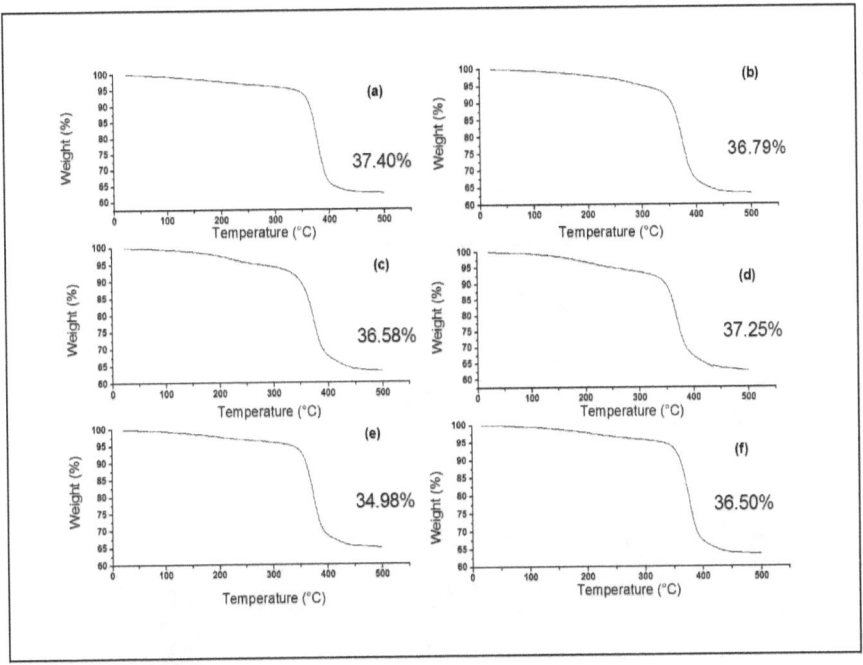

Figure 12

Thermo gravimetric analysis of (a) Acrylic paint (Basic), (b) Commercial antimicrobial acrylic paint, (c) Acrylic paint in presence of ZnO/PD, (d) Acrylic paint in presence of ZnO/T, (e) Acrylic paint in presence of ZnO/DMA and (f) Acrylic paint in presence of ZnO/DPA

3- Nano ZnO and ZnO/amine composites structural features (TEM)

The structural and agglomeration/dispersion features of nano ZnO and its amine composites were studied, results of which are represented in Fig. 13 (a-e). Nano ZnO particles, as prepared is exhibited in Fig. 13 where irregular pattern of particle size prevail at a wide range of 30.9 to 235 nm. The effect of the various amines added proved its role towards ZnO nano particles disintegration differentially. The TEM images revealed that T and PD seem to have a strong function of disintegration on the ZnO particles (15.1-51.2 nm and 27-54 nm), Fig. 13(d,e) with the least disintegration influence represented by DMA and

DPA, Fig. 13(b,c) (56.6 nm and 67.9 nm). The TEM images help in elucidating the role of dispersion of the nano ZnO particles and its effective exposed surface to the biological cells. Amines nature of being aliphatic or aromatic is then regarded as an effective factor towards selectivity and efficiency when evaluating the amine role as an antimicrobial agent. Such concept will be increasingly confirmed by monitoring the biological data of each amine.

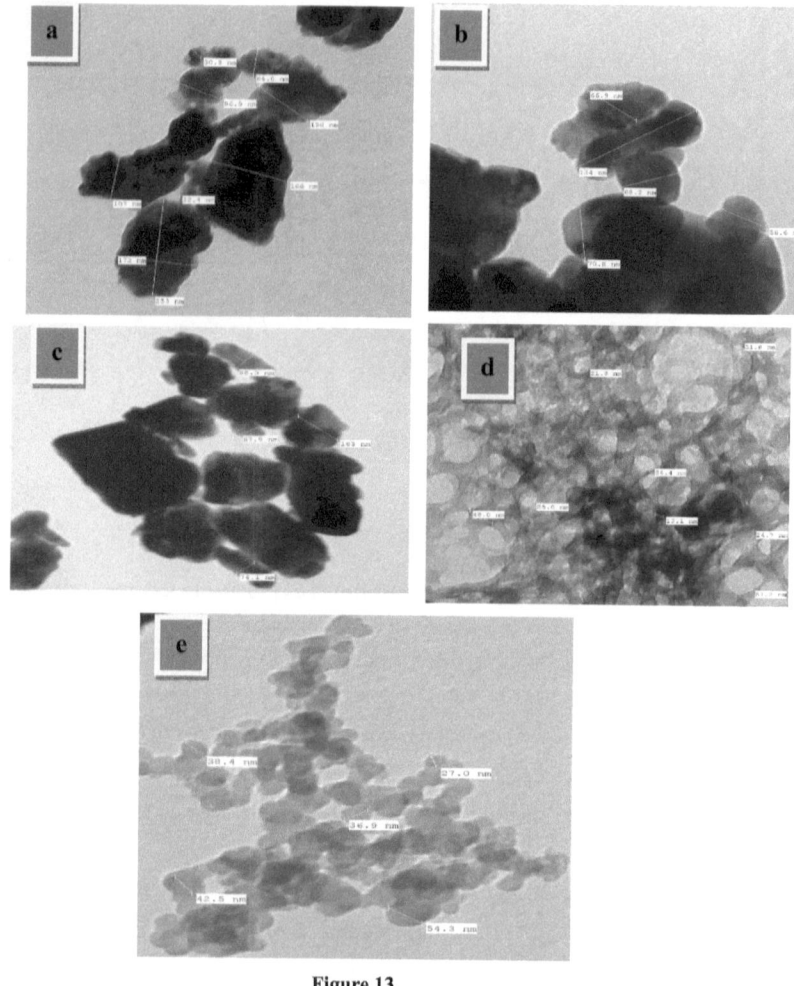

Figure 13
TEM images of (a) nano ZnO, (b) ZnO/DMA, (c) ZnO/DPA, (d) ZnO/T and (e) ZnO/PD.

4- Microbiological activity of native and antimicrobial added acrylic paints

The antimicrobial activity of paint samples, blank and composed nano composites, against the Gram –ve and Gram +ve microorganisms in relation with the particle size of ZnO are illustrated in Table 14. The TEM results illustrate the effect of the additives PD, T, DPA and DMA on the mode of dispersion and agglomeration of nano ZnO particles. PD and T show ability to disperse the nano ZnO particles while DPA and DMA lead to decreasing the dispersion (i.e increase of agglomeration) of nano ZnO (action of surfactancy). Such physical dispersion or agglomeration seems to link directly to the ZnO nano particles activity as an antimicrobial agent. In general, the antimicrobial activity of nano composites of ZnO with PD and T were found higher than those of DPA and DMA which agreed with the concept that ZnO particles with smaller size, larger specific area and higher porosity exhibit higher antimicrobial activity [46].

Higher activity was detected with ZnO/PD acrylic paint resembles than that of the commercially marketed Pachin paint (885), higher than that of pure ZnO. PD acquires specific damaging effect to the peptide links of the cell membrane or the lipids part of the cell wall, due may be to its alcoholic nature structural character, especially with Gram –ve microorganisms that acquire higher lipids content and lower cell wall thickness than those of Gram +ve microorganism. The practical results of turbidity test, Table 14, indicated a higher functional activity of propandiol when ZnO is added which could be assigned to the influence and sustainability of active oxygen species produced by the ZnO environment [49-52] functioning as a specific oxidizing agent to the protein content of the microorganisms cell. The mechanism of the inhibitory effect of ZnO nanoparticles on microorganisms is not fully understood. Several studies reported that integration of ZnO nanoparticles into bacterial cells may induce continuous release of membrane lipids and proteins, which changes the membrane permeability of bacterial cells [48,50]. The combination of ZnO/PD seems an efficient antimicrobial additive due to the dual selective action of both the -diol and the oxide as an overall activity against the protein entity of the microorganism. The antimicrobial activity of triazole, in general, is reported to be referred to the presence of C=N (azomethine group) [50]. When ZnO is added to triazole (ZnO/T), its antimicrobial activity is increased linearly and specifically to Gram +ve microorganisms, Table 11 and Fig. 14 (spots 12, 15).

Table 11

Nano ZnO and its composites, average particle size and respective antimicrobial activity

Additive type	Type of emulsion acrylic paint	ZnO amine composites average particle size (TEM)	Biological activity (Microorganisms resistance) %	
			against Gm -ve	against Gm +ve
Blank	Pachin Virgin (emulsion acrylic paint without additives)	-	0	8
Commercial biocide	Pachin commercial 885 (antimicrobial emulsion acrylic paint)	-	74	98
ZnO	Pachin Virgin emulsion acrylic paint + 3 % nano ZnO	15.9-235	58	36
PD	Pachin Virgin emulsion acrylic paint + 3 % PD	-	80	31
	Pachin Virgin emulsion acrylic paint +3 % (10%ZnO/PD)		84	42
	Pachin Virgin emulsion acrylic paint +3 % (20%ZnO/PD)	14.7 – 71.1	89	55
T	Pachin Virgin emulsion acrylic paint + 3 % T	-	55	14
	Pachin Virgin emulsion acrylic paint + 3 % (10% ZnO/T)		58	19
	Pachin Virgin emulsion acrylic paint + 3 % (20% ZnO/T)	15.1 – 51.2	70	55
DPA	Pachin Virgin emulsion acrylic paint + 3 % DPA	-	42	49
	Pachin Virgin emulsion acrylic paint +3 % (10%ZnO/DPA)		42	49
	Pachin Virgin emulsion acrylic paint +3 % (20%ZnO/DPA)	67.9 - 163	42	49
DMA	Pachin Virgin emulsion acrylic paint + 3 % DMA	-	33	50
	Pachin Virgin emulsion acrylic paint + 3 %	-	64	50

	(10% ZnO/DMA)			
	Pachin Virgin emulsion acrylic paint + 3 % (20%ZnO/DMA)	56.6 – 154	70	50

The following plates, Fig. 14 (3,6,9), illustrate the direct effect of ZnO/T and ZnO/PD on the Gram –ve and +ve microorganisms at 3 % concentration. As shown the ZnO/PD is superior to other amines as an antimicrobial agent with an apparent selectivity towards Gram –ve microorganisms.

DMA is not a strong specific microorganisms assailant by its own, however, the moderate activity for the ZnO/DMA composite could be due to the composite form, Table 14 and Fig. 15 (spots 23 and 26). ZnO /DPA limited antimicrobial activity exhibited, is probably due to limited ability of DPA to bind with the bacteria DNA [51,52] prevailing the ZnO itself even thought the DPA had affected its dispersion negatively by decreasing its exposed active surface to microorganisms decreasing in turn its overall activity compared to bulk ZnO nano particles; Fig. 16.

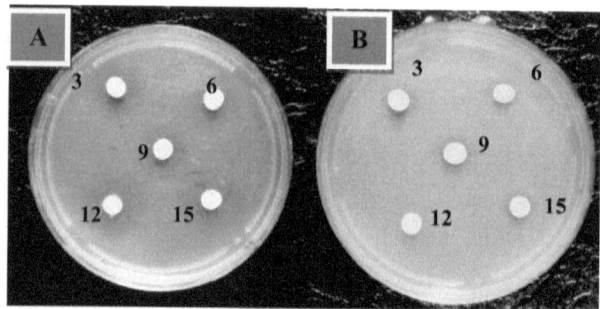

Figure 14

Antimicrobial activities of paints against (A) *Pseudomonas aeruginosa* (gram –ve),
(B) *Staphylococcus aureus (gram +ve)*.
Where 3: the acrylic paint in presence of 3% PD
6: the acrylic paint in presence of 3%(10%ZnO/ PD)
9: the acrylic paint in presence of 3%(20%ZnO/ PD)
12: the acrylic paint in presence of 3% T
15: the acrylic paint in presence of 3%(10%ZnO/ T)f

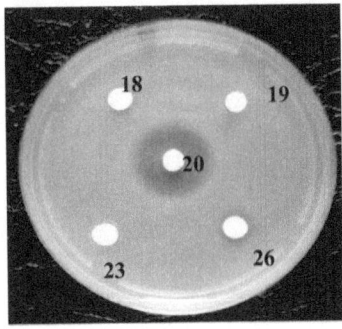

Figure 15

Antimicrobial activity of paints against Staphylococcus aureus

Where 18: the acrylic paint in presence of 3%(10%ZnO/ T)

19:the acrylic paint in absence of biocide (blank)

20: commercial paint of pachin company

23: the acrylic paint in presence of 3%DMA

26: the acrylic paint in presence of 3%(10% ZnO/DMA)

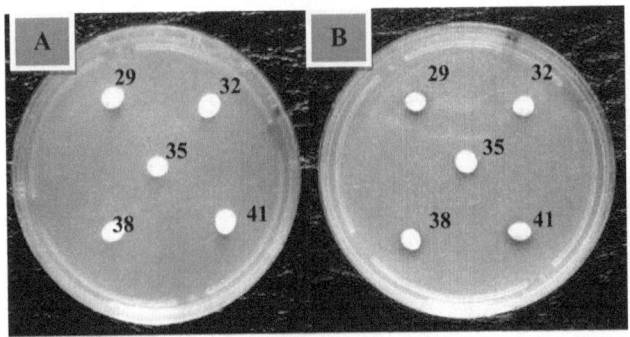

Figure 16

Antimicrobial activities of paints against (A) *Pseudomonas aeruginosa*, (B) *Staphylococcus aureus*.

Where 29: the acrylic paint in presence of 3%(20% ZnO/DMA)

32: the acrylic paint in presence of 3%DPA

35: the acrylic paint in presence of 3%(10%ZnO/ DPA)

38: the acrylic paint in presence of 3% (20%ZnO/ DPA)

41: the acrylic paint in presence of 3% nano ZnO

Rather various cationic surfactants are variably employed as antimicrobial agents as with epoxy paints [53], the present ZnO/amine composites are characterized by its multi attacking mechanisms of both the DNA and the cytoplasmic membrane.

Figures 17 and 18 exhibit a graphical representation of the antimicrobial activity of all additives used in comparison against Gram −ve and Gram +ve microorganisms respectively. Isothiazolone derivatives are being the most effective antimicrobial agent but they are skin irritating and must applied with limit doses which can't used as dry film biocide. It is followed by 20% ZnO/amino PD in case of Gram −ve which higher activity than Pachin biocide that used in the commercial paint but the opposite happened in the case of Gram +ve microorganisms due to weaker effect of amino PD on the thick peptidoglycan layer of the Gram +ve microorganisms cell wall. While the other additives copper and silver complexes in addition to nanocomposites ZnO with DPA and DMA need addition of other agent facilitate the cell wall opening in order to increase their antimicrobial activity.

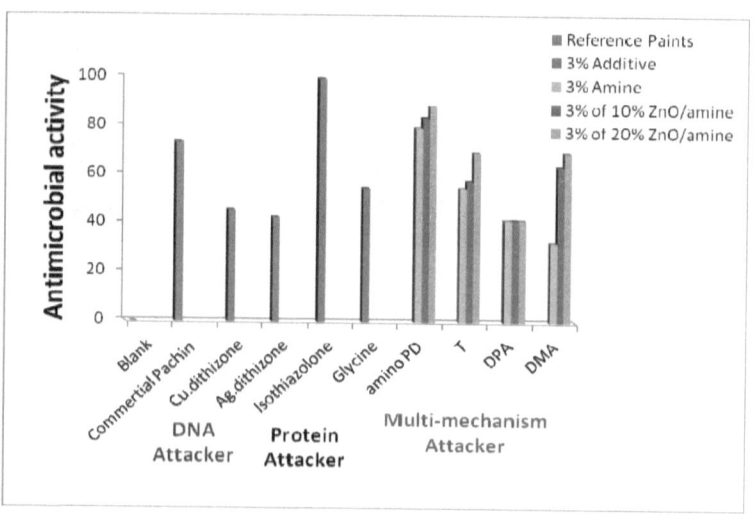

Figure 17

Antimicrobial Activity of Nano Composites of ZnO/amines Against *Pseudomonas aeruginosa* (Gram −ve)

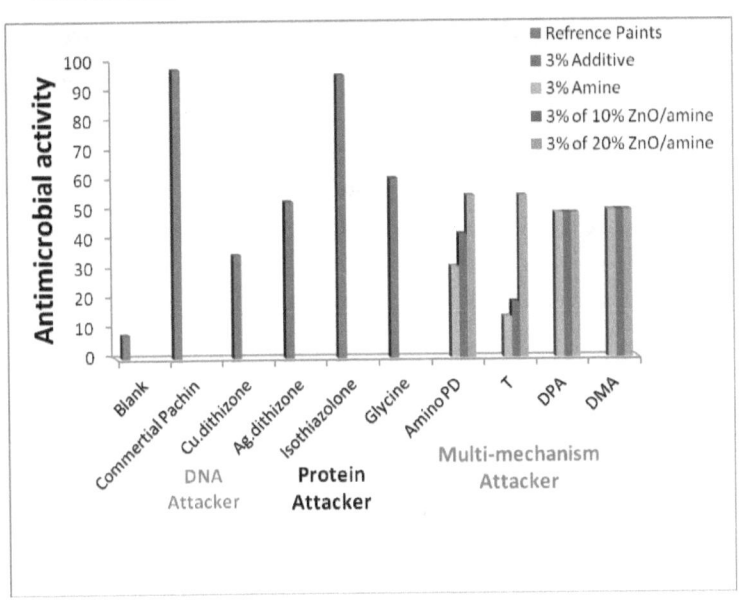

Figure 18

Antimicrobial activity of acrylic paint samples in presence of additives against *Staphylococcus aureus* (Gram +ve)

Part III
Characterization of the three member's additives as in-can preservatives

Cellulose is considered as the main structural component of plant cell walls and is probably the most abundant biological compound on Earth. Cellulose is susceptible to be attacked by a wide range of organisms causing biodeterioration and biofilm formation. Hydroxy ethyl cellulose is the main constituent of emulsion paints as a thickening and soothing agent against coagulation. A microbial enzyme is effective in breaking down the cellulose (natural or synthetic) chains as a direct intake of nutrients. Many organisms excrete waste products, including pigmented or acidic compounds which can disfigure or damage materials causing biodeterioration [54].

A preservative is a substance that is added to products such as foods, pharmaceuticals, paints, biological samples, wood, etc. to prevent decomposition and biological attack by microbial growth or by undesirable chemical changes. The activity of preservatives is due to their antimicrobial activity which prevent degradation by bacteria or due to their *antioxidant properties* that prevent or inhibit the oxidation process and subsequently ripen within the attacked entity [55]. Table 13, illustrates the results of using the three member's group as in-can preservative for emulsion acrylic paints. The virgin acrylic paint tested has no any preservative which is a good compromise for the bacteria and fungi growth, Fig. 19 (A), while the commercial acrylic Pachin paint has a good preservative which remain without rot, Fig. 19 (B). The copper and silver dithizone complexes exhibit antimicrobial activity due to attacking DNA of bacteria [56]. This activity is being more effective in the liquid form of paint.

Kathon biocide (isothiazolone derivatives) is used as preservative in aqueous-polymeric emulsions water-based latex paints, Fig. 20, due to its activity as bactericides and fungicides [57]. Glycine, on the contrary, proved of no antimicrobial activity especially in the liquid form of emulsion paint which could be considered as a nutrient for bacterial growth Fig. 21.

The nanocomposite of ZnO/PD has proved effective as preservative, Fig. 22, inhbiting microorganism growth by modifying the pH level of the emulsion acrylic paint, in addition to its role of sustaining of active oxygen species produced by ZnO which act as oxidizing agent for the protein entity of bacteria [58-61]. The nanocomposite of ZnO/T succeeded herein not as an antimicrobial, antifungal, anticonvulsants and antiviral agent, [62-65] but also as a preservative for the emulsion acrylic paint, Fig. 23. Although the agglomeration of nano ZnO particles was observed as a result of doping with PD, ZnO/DPA, it still retains its antioxidant property [66] and inhibits the oxidation process, Fig. 20.

Table 12, summarizes the results of the additives employed as in-can preservatives differentiating the antimicrobial passed tested samples from those failed to achieve protection against microbial growth.

Table 12

Antimicrobial activity of different compounds when used as preservatives in emulsion acrylic paint

Additive			Result
Local emulsion acrylic paint in absence of in-can preservatives			Failed
In-can preservative additive type			Passed
DNA attacker additives	CuDz. $2H_2O$	0.2%	Passed
		0.3%	Passed
		0.4%	Passed
	AgHDz. H_2O	0.2%	Passed
		0.3%	Passed
		0.4%	Passed
Protein attacker additives	Isothiazolone	0.2%	Passed
		0.3%	Passed
		0.4%	Passed
	Glycine	0.2%	failed
		0.3%	Failed
		0.4%	Failed
Multi-mechanism attacker additives	20% ZnO/PD	0.2%	Passed
		0.3%	Passed
		0.4%	Passed
	20%ZnO/T	0.2%	Passed
		0.3%	Passed
		0.4%	Passed
	20%ZnO/DPA	0.2%	Passed
		0.3%	Passed
		0.4%	Passed
	20%ZnO/DMA	0.2%	failed
		0.3%	failed
		0.4%	failed

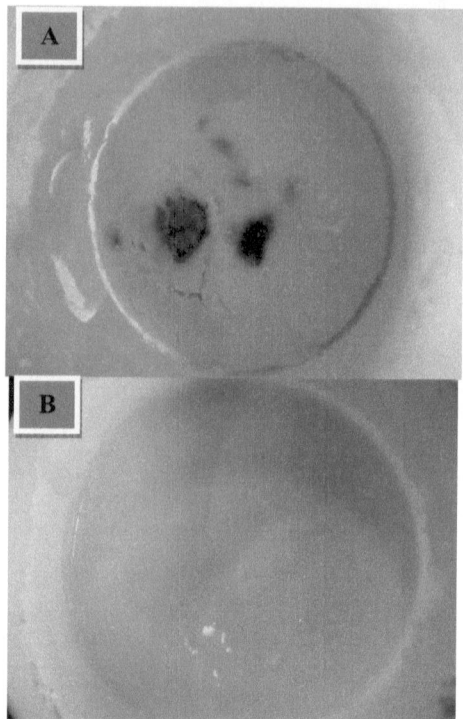

Figure 19

A: Biodeterioration of Emulsion Acrylic Paint (Blank) without Addition of Preservatives.

B: Rot Resistance of Emulsion Acrylic Paint in Presence of Pachin Preservative.

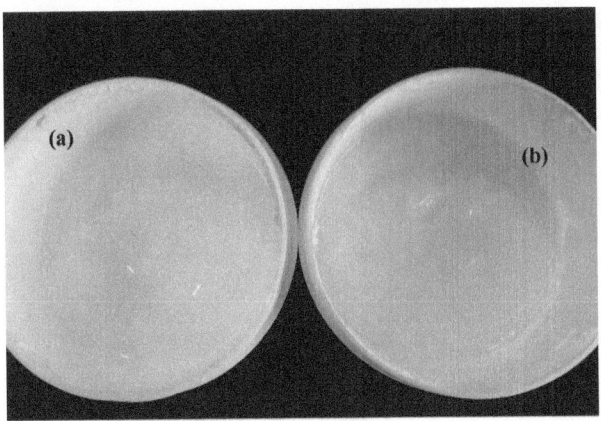

Figure 20

Rot Resistance of Emulsion Acrylic Paint in Presence of (a) 0.4% of 20% ZnO/DPA and (b) 0.4% of Isothiazolone.

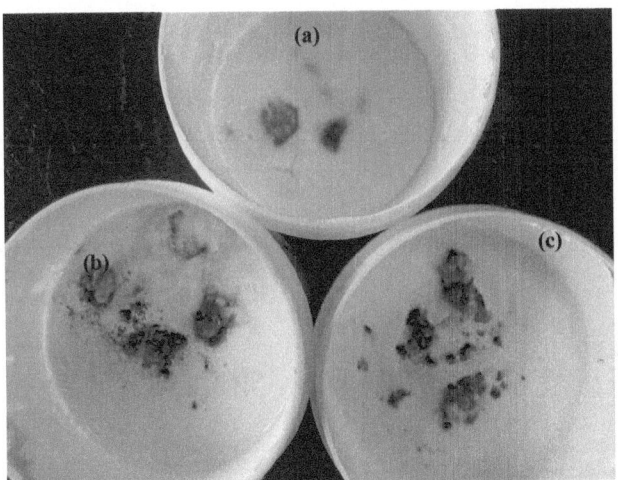

Figure 21

Biodeterioration of emulsion acrylic paint for (a) without addition of preservatives (Blank), (b) in Addition of 0.4% of 20% ZnO/DMA and (c) in addition of 0.4% of glycine.

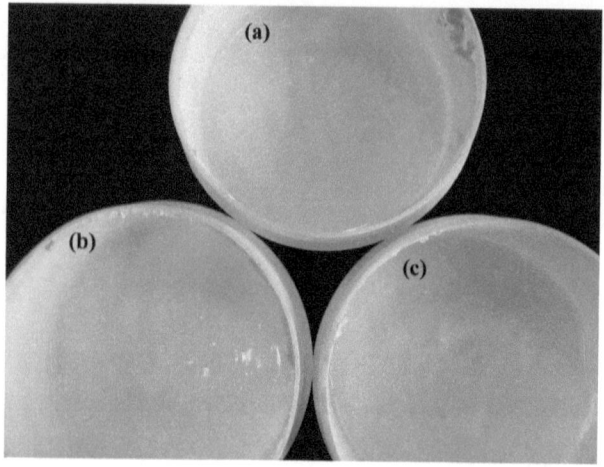

Figure 22
Rot Resistance of Emulsion Acrylic Paint in Presence of (a) 0.2% of 20% ZnO/PD, (b) 0.3% of 20% ZnO/PD and (c) 0.4% of 20% ZnO/PD.

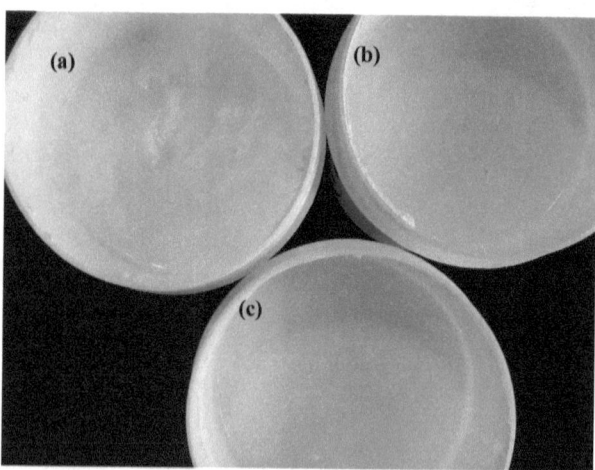

Figure 23
Rot Resistance of Emulsion Acrylic Paint in Presence of (a) 0.2% of 20% ZnO/T, (b) 0.3% of 20% ZnO/T and (c) 0.4% of 20% ZnO/T.

Conclusion

Propandiol, as a more environmental safe additive, is considered as the most suitable amine to composite with ZnO nanoparticles producing an antimicrobial agent equivalent or superseding the commercial available emulsion acrylic paints due to its ability for the production and sustaining the active oxygen species production regarded as responsible for the biological cell damaging. The concluded results indicate the importance of hydroxyl group's presence of the amine additive which might support the active oxygen production sought mechanism when compared to amine groups that might hinder such effect but for a limited period. The TEM images of the nano ZnO particles and its amine composites proved that DPA presence has increased the nano ZnO particles agglomeration leading in turn to a less exposed active sites which could have depreciated its ability as an antimicrobial agent. On the other hand, and while DMA showed fewer particles agglomeration, compared to the DPA, amine aromaticity could be considering as a hindering factor the continuous active oxygen species production responsible for the resistance cause. Rather, triazole (T) illustrated an excessive dispersion of the ZnO particles, elevation of its efficiency as an antimicrobial should be expected, but the depression exhibited could be due the selectivity towards specific bacteria microorganisms compared to the PD amine when functioning in presence of ZnO. Accordingly, it should be concluded that equilibrium between the nano ZnO particles size, their dispersion form, amine ability to stabilize active produced oxygen, type of bacteria microorganisms should all be counted for when persistence of antimicrobial agent efficiency is regarded. In addition to the effective role as antimicrobial coat function the ZnO/amine composites proved to prevail still its function as in-can preservatives even in the presence of cellulose or its derivatives as paints thickening agents.

List of Tables

No.	Title	Page
1	Additives and their colors.	5
2	The composition of paint samples in the presence of different additives	5
3	The mechanical properties of epoxy paint samples	13
4	Thermogravimetric studies (TGA) of epoxy paint samples	17
5	Effect of 5 % NaOH on epoxy paint samples by weight loss technique	19
6	Effect of 10% HCl acid on epoxy paint samples by weight loss technique	20
7	Effect of artificial sea water on epoxy paint samples by weight loss technique	21
8	Diameter of inhibition zones of different antibiotics (cm) against 12 clinical isolates	25
9	FT-IR absorption spectral bands (cm-1) of ZnO, amine additives and ZnO composites characterization	34
10	FT-IR spectra analysis (cm-1) of basic poly acrylic paint and its ZnO/amine added characterization	35
11	Nano ZnO and its composites, average particle size and respective antimicrobial activity	40
12	Antimicrobial activity of different compounds when used as preservatives in emulsion acrylic paint	46

List of Figures

No.	Title	Page
1	FT-IR spectra of (a) epoxy resin, (b) hardener and (c) native epoxy paint after curing.	10
2	FT-IR spectra of (a) DAS, (b) epoxy paint in presence of DAS:$CoCl_2$.	10
3	FT-IR spectra of (a) Hg $(HDz)_2$, (b) epoxy paint in presence of Hg $(HDz)_2$.	12
4	Thermo-gravimetric analysis of (a) native epoxy paint, (b) epoxy paint in presence of DAS:$CoCl_2$ and (c) epoxy paint in presence of (Hg $(HDz)_2$).	16
5	Relationship between the weight loss % of immersed paint samples in (5% NaOH) and immersion time for alkali resistance.	19
6	Relationship between the weight loss % of immersed paint samples in (10% HCl) and immersion time for acid resistance.	20
7	Relationship between the weight loss% of immersed paint samples and immersion time for artificial sea water resistance.	21
8	The photo stability of Hg $(HDz)_2$ complex solution.	23
9	The photo stability of epoxy paint modified by Hg $(HDz)_2$.	24
10	Antimicrobial activities of paint samples against 12 clinical isolates. III, paint disc impregnated with Hg $(HDz)_2$, II, paint disc impregnated with DAS:$CoCl_2$.	25
11	Antimicrobial activities of paints against A) Staphylococcus aureus Sa1, B) Staphylococcus aureus Sa2 and C) Pseudomonas aeruginosa Pa1. III, paint disc impregnated with Hg $(HDz)_2$, II, paint disc impregnated with DAS:$CoCl_2$.	26
12	Thermo gravimetric analysis of (a) Acrylic paint (Basic), (b) Commercial antimicrobial acrylic paint, (c) Acrylic paint in presence of ZnO/PD, (d) Acrylic paint in presence of ZnO/T, (e) Acrylic paint in presence of ZnO/DMA and (f) Acrylic paint in presence of ZnO/DPA.	36
13	TEM images of (a) nano ZnO, (b) ZnO/DMA, (c) ZnO/DPA, (d) ZnO/T and (e) ZnO/PD.	37
14	Antimicrobial activities of paints against (A) Pseudomonas aeruginosa (gram –ve), (B) Staphylococcus aureus (gram +ve). Where 3: the acrylic paint in presence of 3% PD 6: the acrylic paint in presence of 3%(10%ZnO/ PD) 9: the acrylic paint in presence of 3%(20%ZnO/ PD) 12: the acrylic paint in presence of 3% T 15: the acrylic paint in presence of 3%(10%ZnO/ T)	41
15	Antimicrobial activity of paints against Staphylococcus aureus	42

16	Where 18: the acrylic paint in presence of 3%(10%ZnO/ T) 19: the acrylic paint in absence of biocide (blank) 20: commercial paint of pachin company 23: the acrylic paint in presence of 3%DMA 26: the acrylic paint in presence of 3%(10% ZnO/DMA) Antimicrobial activities of paints against (A) Pseudomonas aeruginosa, (B) Staphylococcus aureus. Where 29: the acrylic paint in presence of 3%(20% ZnO/DMA) 32: the acrylic paint in presence of 3%DPA 35: the acrylic paint in presence of 3%(10%ZnO/ DPA) 38: the acrylic paint in presence of 3% (20%ZnO/ DPA) 41: the acrylic paint in presence of 3% nano ZnO	42
17	Antimicrobial Activity of Nano Composites of ZnO/amines Against *Pseudomonas aeruginosa* (Gram –ve)	43
18	Antimicrobial activity of acrylic paint samples in presence of additives against *Staphylococcus aureus* (Gram +ve)	44
19	(A) Biodeterioration of Emulsion Acrylic Paint (Blank) without Addition of Preservatives and (B) Rot Resistance of Emulsion Acrylic Paint in Presence of Pachin Preservative.	47
20	Rot Resistance of Emulsion Acrylic Paint in Presence of (a) 0.4% of 20% ZnO/DPA and (b) 0.4% of Isothiazolone.	48
21	Biodeterioration of emulsion acrylic paint for (a) without addition of preservatives (Blank), (b) in Addition of 0.4% of 20% ZnO/DMA and (c) in addition of 0.4% of glycine.	48
22	Rot Resistance of Emulsion Acrylic Paint in Presence of (a) 0.2% of 20% ZnO/PD, (b) 0.3% of 20% ZnO/PD and (c) 0.4% of 20% ZnO/PD.	49
23	Rot Resistance of Emulsion Acrylic Paint in Presence of (a) 0.2% of 20% ZnO/T, (b) 0.3% of 20% ZnO/T and (c) 0.4% of 20% ZnO/T.	49

References

[1] Ahmed, S.; Gupta, A. P.; Sharmin, E.; Alam, M. and Pandey, S. K. (2005) Prog. Org. Coat. 54:248.

[2] Chae, K.H.; Jang, Y.M.; Kim, Y.H.; Sohn, O. and Rhee, J. (2007) Sensors and Actuators B 124:153.

[3] Aggarwal, L. K.; Thapliyal, P. C. and Karade, S. R. (2007) Prog. Org. Coat. 59:76.

[4] Atta, A. M.; Shaker, N. O. and Maysour, N. E. (2006) Prog. Org. Coat. 56:100.

[5] Roche, A. A.; Bouchet, J. and Bentadjine, S. (2002) International Journal of Adhesion & adhesives 22:431.

[6] Rajesh, N. and Hari, M.S. (2008) Spectrochimica Acta Part A 70:1104.

[7] Balgavf, P. and Devinsky, F. (1996) Advances in Colloid and Interface Science 66:23.

[8] Teshima, K.; Yoneyama, T. and Kondo, T. (2008) Journal of Pharmaceutical and Biomedical Analysis 47:962.

[9] Blount, D.H.; Jul. 11 (1978) United States Patent of ser No 798,329.

[10] Meriwether, L.S.; Breitner, E.C. and Sioan, C.L. (1965) American Chemical Society 87:4441.

[11] Atta, A.M.; Mansour, R.; Abdou, M.I. and Sayed, A.M. (2004) Polym. Adv. Technol. 15:514.

[12] Kumar, S. A.; Balakrishan, T.; Alagar M.; and Denchev, Z. (2006) Prog. Org. Coat. 55:207.

[13] Cheesbrough,M. (1989) Medical Laboratory Manual for Tropical Countries, Vol 2: Microbiology. Tropical Health Technology/Butter- worth and Co. Ltd. Kent.

[14] Adonizio, A.L.; Downum, K.; Bennett B.C. and Mathee, K. (2006) J. Ethnopharmacol. 105:427.

[15] Tait, J.K.F.; Davies, G.; Mcintyre, R. and Yarwood, J. (1997) Vibrational Spectroscopy 15:79.

[16] Bajpai, M; Shukla, V. and Habib F. (2005) Prog. Org. Coat. 53:239.

[17] Hamdy, A. (2006) Prog. Org. Coat. 55:218.

[18] Meriwether, L.S.; Breitner, E.C. and Clothup, N.B. (1965) J. Am. Chem. Soc. 87:4448.

[19] Chlebicki, J.; Wegrzynska, J.; Maliszewska, I.; and Oswiecimsk, M. (2005) J. Surfact. And Deterg. 8:227.

[20] Hamilton, W.A. (1968) J. Gen. Microbiol., 50: 441.

[21] Ioannou, C. J.; Hanlon, G. W. and Denyer, S. P. (2006) Antimicrob. Agents Chemother. 51:296.

[22] Rauter, A.P.; Lucas, S.; Almeida, T.; Sacoto, D.; Ribeiro, V.; Justino, J.; Neves, A.; Silva, F.V.M.; Oliveira, M.C.; Ferreira, M.J.; Santos, M. and Barbosa, E. (2005) Carbohydrate Research 340: 191.

[24] J. Pasquet, Y. Chevalier, E. Couval, D. Bouvier, G. Noizet, C. Morlière, M. Bolzinger, International Journal of Pharmaceutics 460 (2014) 92-100.

[25] A. Akbar, A. Kumar, Anal. Food Control 38 (2014) 88-95.

[26] K. M. Kumar, B. K. Mandal, E. A. Naidu, M. Sinha, K. S. Kumar, P. S. Reddy, Spectrochimica Acta Part A: Molecular and Biomolecular Spectroscopy 104 (2013) 171-174.

[27] N. Jones, B. Ray, K. T. Ranjit, A. C. Manna, FEMS Microbiol. Lett. 2 (2008) 71-76.

[28] Y. Liu, L. He, A. Mustapha, H. Li, , Z.Q. Hu, M. Lin, J. Appl. Microbiol. 107 (2009) 1193-1201.

[29] R. Brayner, R. Ferrari-Iliou, N. Brivois, S. Djediat, M. F. Benedetti, F. Fie´vet, Nano Lett. 6 (2006) 866-870.

[30] J. Sawai, J. Microbiol. Methods 54(2003) 177-182.

[31] P. K. Stoimenov, R. L. Klinger, G. L. Marchin, K. J. Klabunde, Langmuir, 18 (2002) 6679-6686.

[32] O. Yamamoto, M. Komatsu, J. Sawai, Z. E. Nakagawa, J. Mater Sci. Mater. Med., 15 (2004) 847-851.

[33] A. Tayel, W. A,El-Tras, , S. Moussa, , M. Salemi, , L. Brimer, , J. Food Safety 31 (2011) 211-218.

[34] M. Mekewi, A. Shebl, I.A. Imam, M. S. Amin, T. Albert, J. Materials Science & Technology 28 (2012) 961–968.

[35] M. Mekewi, A. A. El-Sayed, M. S. Amin, H. I. Saied, International Journal of Biological Macromolecules, 50 (2012) 1055-1062.

[36] C. Guran, A. Pica, D. Fical, A. Fical, C. Comanescu, Bul. Mat. Sci., 13 (2013) 183-188.

[37] T. K. Sontakke, S. Jagtap, D. C. Kothari, Progress in Organic Coatings 74 (2012) 582-588.

[38] M. Kathalewara, A. Sabnisa, G. Waghoo; Progress in Organic Coatings 76 (2013) 1215– 1229

[39] M. Cheesbrough, Tropical Health Technology/Butter- worth and Co. Ltd. Kent (1989).

[40] A. L. Adonizio, K. Downum, B.C. Bennett, K. Mathee; J. Ethnopharmacol., 105 (2006) 427–435.

[41] ö. Topçuoğlu, S. A. Altinkaya, D. Balköse, Progress in Organic Coatings 56 (2006) 269–278.

[42] S. M. Fufa, B. P. Jelle, P. J. Hovde, Progress in Organic Coatings 76 (2013) 1543– 1548

[43] X.W. Du, Y.S. Fu, J. Sun, X. Han, J. Liu, Semicond. Sci. Technol. 21 (2006) 1202–1206.

[44] S. A. Ansari , Q. Husain, S. Qayyum, A. Azam, Food and Chemical Toxicology 49 (2011) 2107–2115.

[45] R. John Xavier, E. Gobinath, Spectrochimica Acta Part A 86 (2012) 242–251.

[46] R. Almeida, A. Gómez-Zavaglia, A. Kaczor, A. Ismael, M. L. S. Cristiano, R. Fausto, Journal of Molecular Structure 938 (2009) 198–206.

[47] R. Rajamohan, M. Swaminathan, Spectrochimica Acta Part A 83 (2011) 207– 212.

[48] M. Jia, K. Yang, H. Fang, Y. Xu, S. Sun, L. Su, W. Xu, Bioorganic & Medicinal Chemistry 19 (2011) 5190–5198.

[49] N. A. Amro, L. P. Kotra, K. Wadu-Mesthrige, A. Bulychev, S. Mobashery, G. Liu, Langmuir 16 (2000) 2789—2796.

[50] A. M. Vijesh, A. M. Isloor, P. Shetty, S. Sundershan, H. Kun Fun, European Journal of Medicinal Chemistry 62 (2013) 410-415.

[51]T. Pederson, Analytical Biochemistry 28 (1969) 35-46.

[52] X. Li, Y. Wu, L. Zhang, Y. Cao, Y. Li, J. Li, L. Zhu, G. Wu, Analytical Biochemistry 451(2014) 18-24.

[53] M.S. Antonious, A.F. Badawi, M.A.Mekewi H.B.Kamal, Ain-Shams Science Bulletin, 46, (2008) 51-66

[54] Dennis Allsopp, Kenneth J. Sea and Christine C. Gaylarde; (Introduction to Biodeterioration) Cambridge University Press. United Kingdom. Second Edition ISBN 0-521-82135-5-ISBN 0-521-52887-9 (pb) (2004)

[55] Erich Lück and Gert-Wolfhard von Rymon Lipinski "Foods, 3. Food Additives" in Ullmann's Encyclopedia of Industrial Chemistry, 2002, Wiley-VCH, Weinheim. doi: 10.1002/14356007.a11_561

[56] J. Kohroki, N. Muto, T. Tanaka, N. Itoh, A. Inada and K. Tanaka; Leukemia research 22 (1998) 405-412

[57] Rohm and Haas; US Patent No. 3761488/1973.

[58] Jones, N., B. Ray, K. T. Ranjit, and A. C. Manna; FEMS Microbiol. Lett. 2 (2008) (79) 71-76.

[59] Liu, Y., He, L., Mustapha, A., Li, H., Hu, Z.Q. and Lin, M.; J. Appl. Microbiol. 107 (2009) 1193-1201.

[60] Roberta Brayner, Roselyne Ferrari-Iliou, Nicolas Brivois, Shakib Djediat, Marc F. Benedetti, and Fernand Fie'vet; Nano Lett. 6 (2006) 866-870.

[61] Sawai, J.; J. Microbiol. Methods, 54 (2003) 177-182.

[62] R.M. McCarrick, M.J. Eltzroth, Ph.J. Squattrito, Inorg. Chim. Acta 311 (2000) 95-105.

[63] O. Bekircan, N. Gümrükçüo_glu, Indian J. Chem. 44B (2005) 2107-2113.

[64] N. Eweiss, A. Bahajaj, E. Elsherbini, Heterocycl. Chem. 23 (1986) 1451-1458.

[65] I. Awad, A. Abdel-Rahman, E. Bakite, J. Chem. Technol. Biotechnol. 51 (1991) 483-495.

[66] Takashi Sugihara, Gundo Rao and Robert P. Hebbel; Free Radical Biology and Medicine 14 (1993) 381-387.

Acknowledgement

The authors would like to extend their appreciation to Ain Shams University for their research grant which enabled financing the present work within the goals of scientific research environmental motivations programs. The authors would also like to acknowledge the assistant given by the local paints manufacturers for their support and motivations.

i want morebooks!

Buy your books fast and straightforward online - at one of the world's fastest growing online book stores! Environmentally sound due to Print-on-Demand technologies.

Buy your books online at
www.get-morebooks.com

Kaufen Sie Ihre Bücher schnell und unkompliziert online – auf einer der am schnellsten wachsenden Buchhandelsplattformen weltweit!
Dank Print-On-Demand umwelt- und ressourcenschonend produziert.

Bücher schneller online kaufen
www.morebooks.de

OmniScriptum Marketing DEU GmbH
Heinrich-Böcking-Str. 6-8
D - 66121 Saarbrücken
Telefax: +49 681 93 81 567-9

info@omniscriptum.de
www.omniscriptum.de

www.ingramcontent.com/pod-product-compliance
Lightning Source LLC
Chambersburg PA
CBHW031545210526
45464CB00003B/1159